HANDBOOK FOR INDUSTRIAL NOISE CONTROL

Submitted to
National Aeronautics and Space Administration
Langley Research Center
Hampton, Virginia

Prepared under Contract NAS1-15721
by
The Bionetics Corporation
Hampton, Virginia

Books for Business
New York - Hong Kong

Handbook for Industrial Noise Control

Submitted to
National Aeronautics and Space Administration

by
The Bionetics Corporation

ISBN: 0-89499-065-9

Preface

This handbook gives basic and comprehensive information on the understanding, measurement, and control of noise in industrial environments. It is intended for engineers with or without acoustical experience; to this end, it presents sections on noise problem analysis, instrumentation, fundamental methods of noise control, and properties of acoustical materials.

The National Aeronautics and Space Administration has been one of the chief instigators of many important research and development studies within the field of aeroacoustics. These have been conducted in-house, by major aerospace contractors to NASA, and by academic institutions under the aegis of the extensive NASA research grant program. The material included in this book is limited to that which is clearly applicable to nonaerospace industrial noise control problems. The later chapters, in particular, include more advanced and source-specific noise control technology. Emphasis has been placed on fan noise reduction, noise transmission control techniques, and jet noise suppression.

An extensive bibliography and reference list of books and articles has been assembled. Many of these resulted from NASA sponsorship, and, to a lesser extent, that of other federal agencies.

The author makes no claim to originality regarding the basic factual content of this work. In blending some of the more esoteric NASA-sponsored research with fundamental, well-established tenets of noise control, a careful selectivity must be exercised. In preparing this work, many NASA formal publications and journal articles were reviewed. The author is solely responsible for the material included here, and he apologizes to those who may feel that their work has been overlooked.

W. Graham Orr
Hampton, Virginia

Acknowledgment

This document was prepared by The Bionetics Corporation for NASA under contract No. NAS1-15721. The author thanks Kevin P. Shepherd, Ph.D., Project Manager, for his patient assistance in proofreading the text and for his technical counsel. He also extends his sincere appreciation to Lester J. Rose of NASA Langley Research Center, Technical Representative of the Contracting Officer, for advice and coordination of review.

Too numerous to mention are the many individuals and companies who responded to requests for information by providing updates of research work in progress, published reports, and technical specifications of equipment and materials. Source credits are given next to all photographs that appear in the text.

Finally, the support of the staff at Bionetics is acknowledged: Alan J. Rosing for his judicious technical editing; Sandy Barnes for her splendid artwork; and Frankie Freeze, Michael Platt, and Nancy Walton for the exceptional care taken in typing the manuscript.

Contents

CHAPTER 1

Introduction

Noise is commonly described as unwanted sound. This definition implies a strong subjective element in any assessment of the effects of noise, since what may be tolerated by one person may be intolerable to another. Although noise is by no means a recent phenomenon, it is only with the rapid industrialization of the last century and the concomitant accelerated development of powerful high-speed machinery that it has become a pollutant of major concern. The effects of noise exposure on humans range from disturbance or annoyance to temporary or even irreversible deafness.

Conventionally, industrial noise control is aimed at alleviating conditions likely to lead to hearing impairment. In 1969 the first federal regulation limiting civilian occupational noise exposure was promulgated under the Walsh-Healey Public Contracts Act of 1936. Initially these limits applied only to certain government suppliers, but they were soon extended to all industries by regulations under the Occupational Safety and Health Act of 1970. These standards remain the most pertinent to industrial occupational noise control and are given in detail in the appendix.

Currently, 90 dB(A) is the maximum allowable continuous noise level that can occur throughout an 8-hour day. At higher levels some form of ameliorating action must be taken. There has been considerable pressure from organized labor, NIOSH,[1] and the Environmental Protection Agency[2] to reduce this limit to 85 dB(A) or less, but serious questions concerning the benefits and costs of such a reduction are unresolved.

It is estimated that some 3.5 million Americans now work in environments in which the OSHA noise standards are exceeded, and that 13.5 million risk some form of hearing loss. Even with a stricter standard than that currently in force, a proportion of workers will be at risk because of variations in individual susceptibility to hearing damage. One study[4] estimates that a total capital cost of $8 billion would be required to achieve engineering compliance with a standard of 85 dB(A).

An aspect of industrial noise that is not considered in detail here is the effect of noise levels outside the work place on the community and environment. The Noise Control Act of 1972 governs such noise levels and provides for regional enforcement through state and local noise ordinances.[5] One of the main provisions of the Noise Control Act is the development of Federal noise emission standards for major noise sources in the categories of construction, transportation, motors and engines, and electrical or electronic equipment in commerce. The EPA is responsible for enforcing this act and is required to develop standards and suitable labeling of these noise sources, as well as for such noise control devices as hearing protectors. The act, together with the extensive provisions of the Quiet Communities Act of 1978, gives EPA a charter to conduct and finance noise control research. The bibliography at the end of this book includes several important reports published as a result of such work. The EPA is also empowered to act as the chief coordinator of all Federal agency noise control programs. The EPA has proposed that community noise levels be reduced to $L_{dn} = 65$ dB(A) as soon as possible[2] and will enlist the help of state and local governments toward this objective.

Partly as a result of research authorized by the Noise Control Act, increased knowledge concerning the nonauditory effects of noise is being sought. These effects may arise from nonadaptive physiological responses to noise as a nonspecific biological stressor.[6, 7] Some of the more important responses may be increased cardiovascular disease, elevated blood cholesterol levels, increased adrenal hormone production, blood vessel constriction, and accelerated heartbeat.[7] Although no firm scientific consensus has emerged on the possible link between noise and these conditions, the potential benefits of noise reduction are certainly not confined to the reduction of hearing damage.

Chapters 2–4 are a comprehensive introduction to the basic physical principles, measuring techniques, and instrumentation associated with general purpose noise control. Chapter 3 includes an outline showing how best to identify and characterize a noise problem so that subsequent work may provide the most efficient and cost-effective solution. A detailed methodology for choosing appropriate noise control materials and the proper implementation of control procedures begins in chapter 5. The most significant NASA-sponsored contributions to the state-of-the-art development of optimum noise control technologies are contained in the final chapter. Of particular interest are cases in which aeroacoustics and related research have shed some light on ways of reducing noise generation at its source.

Equations and associated mathematics are presented in a simple, comprehensible manner. To aid the reader, examples are presented at pertinent points in the text. In addition, there is a substantial categorization of the text and a large selection of figures illustrating particular items of equipment, nomograms, and plots of performance characteristics.

Although it is not explicitly stated, many of the references, especially those listed at the end of the later chapters, resulted from NASA-sponsored research. This enables such material to be placed within a relevant and proper context.

REFERENCES

1. NIOSH, Criteria for a Recommended Standard—Occupational Exposure to Noise. HSM 73-11001, 1972.
2. Office of Noise Abatement and Control, EPA, Toward a National Strategy for Noise Control, April 1977.
3. Office of Noise Abatement and Control, EPA, Noise Technology Research Needs and the Relative Roles of the Government and the Private Sector, EPA 550/9-79-311, May 1979.
4. Bruce, R. D., et al., Economic Impact Analysis of Proposed Noise Control Regulation, BBN Report No. 3246, April 1976.
5. Haag, F. G., State Noise Restrictions: 1979, Sound and Vibration, December 1979, p. 16.
6. Elkins, C., Noise: The Invisible Pollutant, EPA Journal, October 1979.
7. Center for Policy Alternatives, MIT (prepared for EPA), Some Considerations in Choosing an Occupational Noise Exposure Regulation. EPA 550/9-76-007, February 1976.

CHAPTER 2

Basic Physics of Sound

Sound is a mechanical disturbance of a wave nature that propagates through any elastic medium at a speed characteristic of that medium. When detected by the human ear, the presence of sound waves is often described by the subjective attribute of loudness, and this quality is usually quantified by averaging over a range of people (see section 3.1.3). Loudness is related to the amplitude of movement in the transmitting medium next to the eardrum.

Sound may be assessed after measurement of an appropriate physical quantity that varies about an equilibrium position as the sound wave or vibration passes the measuring point. Sound pressure is the macroscopic physical quantity of most interest in fluids and gases. It depends on the amplitude of the disturbance in a material and most adequately represents the effect of acoustic waves on human response. Figure 2.1 illustrates how an acoustic pressure disturbance might vary about the ambient value for a typical sound wave. Atmospheric pressure is taken as the base line for airborne sound propagation.

As an illustration of the magnitude of the fluctuations involved, atmospheric pressure is normally 1.012×10^5 N/m^2 at sea level; a barely audible signal might fluctuate as little as 10^{-5} N/m^2 from this value, and a painfully loud continuous noise only about 10 N/m^2, that is, only 0.01% of the atmospheric value. (The unit of newtons per square meter, N/m^2, is sometimes replaced by its equivalent, the pascal, Pa.)

2.1 SIMPLE WAVE NATURE OF SOUND

2.1.1 Introduction to Longitudinal Propagation

Wave motion or vibration is made possible by inertia and elasticity in a material. Because of inertia, matter remains at rest or in uniform motion in the absence of external forces; this permits the transfer of momentum between neighboring particles or elements in a medium. Elasticity designates the tendency of displaced elements to return to their original or equilibrium position after the passage of a disturbance. Air, water, glass, and such metals as steel and iron are relatively elastic.

Wave propagation may be demonstrated by the use of a long helical spring fixed at one end. If a disturbance is imparted at the free end, the motion is carried through the adjacent coils, while at the same time the initially displaced coils tend to return to their former positions. The net effect is to produce two distinct regions that travel down the coil structure: in one, a series of neighboring coils are pushed close together because of the initial action of the disturbance (compression); in the other, the relative spacing of coils is enlarged because of the removal of the disturbing force and subsequent overcompensation by the elastic coils (rarefaction).

The wave nature of sound in air may also be demonstrated by the device presented in figure 2.2.

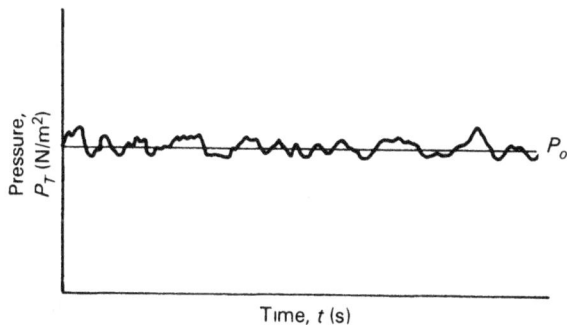

Figure 2.1 - Example of acoustic pressure fluctuations about mean ambient pressure. P_o is the ambient atmospheric pressure; p is the acoustic pressure; and P_T is the total pressure (equal to $P_o + P_T$).

3

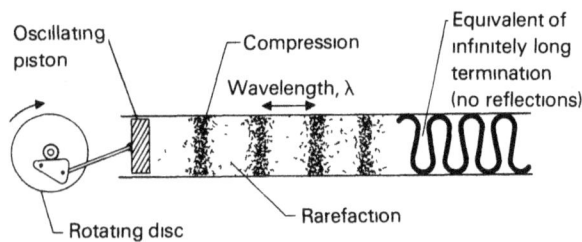

Figure 2.2 - *Generation of longitudinal progressive waves in a tube.*

The action of the piston in figure 2.2 alternately pushes the air molecules together or pulls them apart. It is the inherent elasticity of air, combined with inertia, that allows these successive disturbances to travel down the tube as the displaced molecules try to regain their former equilibrium. The net effect on the pressure distribution is maximum pressure amplitude at points of maximum compression and minimum pressure at points of maximum rarefaction. If the disc rotates at a constant speed, then the wave form, as represented by the pressure, will be sinusoidal. Figure 2.3 shows some of the characteristics of this simple wave form, termed a longitudinal plane progressive wave because the elements of the medium are displaced in the direction of propagation and the wave is not stationary in space. Waves that propagate in many duct noise control problems or at large distances in free space from a spherically radiating source closely approximate longitudinal plane progressive waves.

2.1.2 Frequency

The representation of pressure change at a point in space with time, $p(t)$, or at a fixed instant of time throughout space, $p(x)$, as shown in figure 2.3, is described by the equations

$$p(t) = P_0 \sin 2\pi f t$$
$$p(x) = P_0 \sin 2\pi f x$$

where

P_0 = equilibrium or ambient sound pressure (N/m^2)
f = frequency of the disturbance (Hz)
x = distance (m)
t = time (s)

A complete cycle of sound pressure oscillation is normally represented by 360° or 2π radians. The symbol commonly used to denote frequency is f, which is the number of complete cycles in a second. The fundamental unit of measurement is cycles per second or Hertz (Hz). Sometimes the term "period" is invoked to describe the nature of the wave. This is simply the time taken to complete one cycle, T; it is equal to the reciprocal of the frequency.

The normal frequency range of adult hearing extends from about 20 to 20 000 Hz. Because of its innate characteristics, the human ear is most sensitive to sound around 3 kHz, especially at the threshold of audibility. The subjective attribute known as pitch, which is often used to describe musical quality, is highly correlated with frequency. The sound waves normally associated with noise in industry are very rarely single frequencies in isolation (also known as pure tones). In general, there are two classes of sounds: those which repeat at regular intervals (periodic) and most often consist of a variety of related frequencies occurring together, and those which fluctuate randomly and do not repeat (aperiodic). The latter sounds may be considered part of an infinitely long period.

A mathematical technique known as Fourier analysis allows any periodic signal to be partitioned into a series of harmonically related pure tone signals. This enables any complex but regular waveform to be described in terms of a number of discrete frequencies, each with a particular amplitude. A pictorial representation of this is

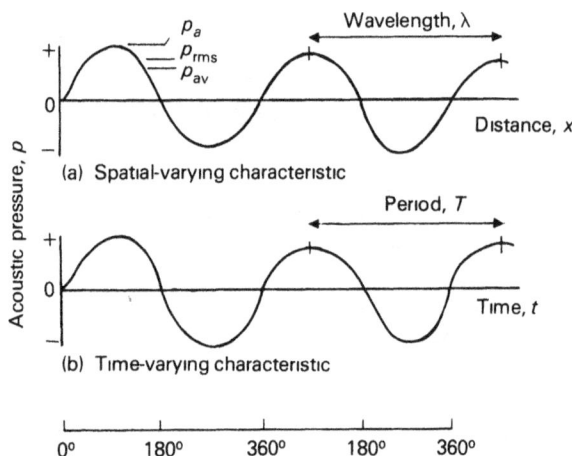

where
p_{rms} = root mean square sound pressure
p_{av} = average sound pressure during positive half-cycle
p_a = maximum sound pressure amplitude

Figure 2.3 - *Characteristics of simple acoustic sine waves.*

4

given in figure 2.4a, showing how three superimposed sinusoidal waveforms can be used to produce a more complex waveform. The presence of a harmonic relationship means that the frequency of each wave is related to the fundamental or lowest frequency by an integer multiple. A frequency spectrum simply describes the relative amplitude of each single frequency component; these are shown in figure 2.4b, corresponding to the various waveforms in figure 2.4a.

In contrast, aperiodic sounds can only be treated as consisting of an infinitely large number

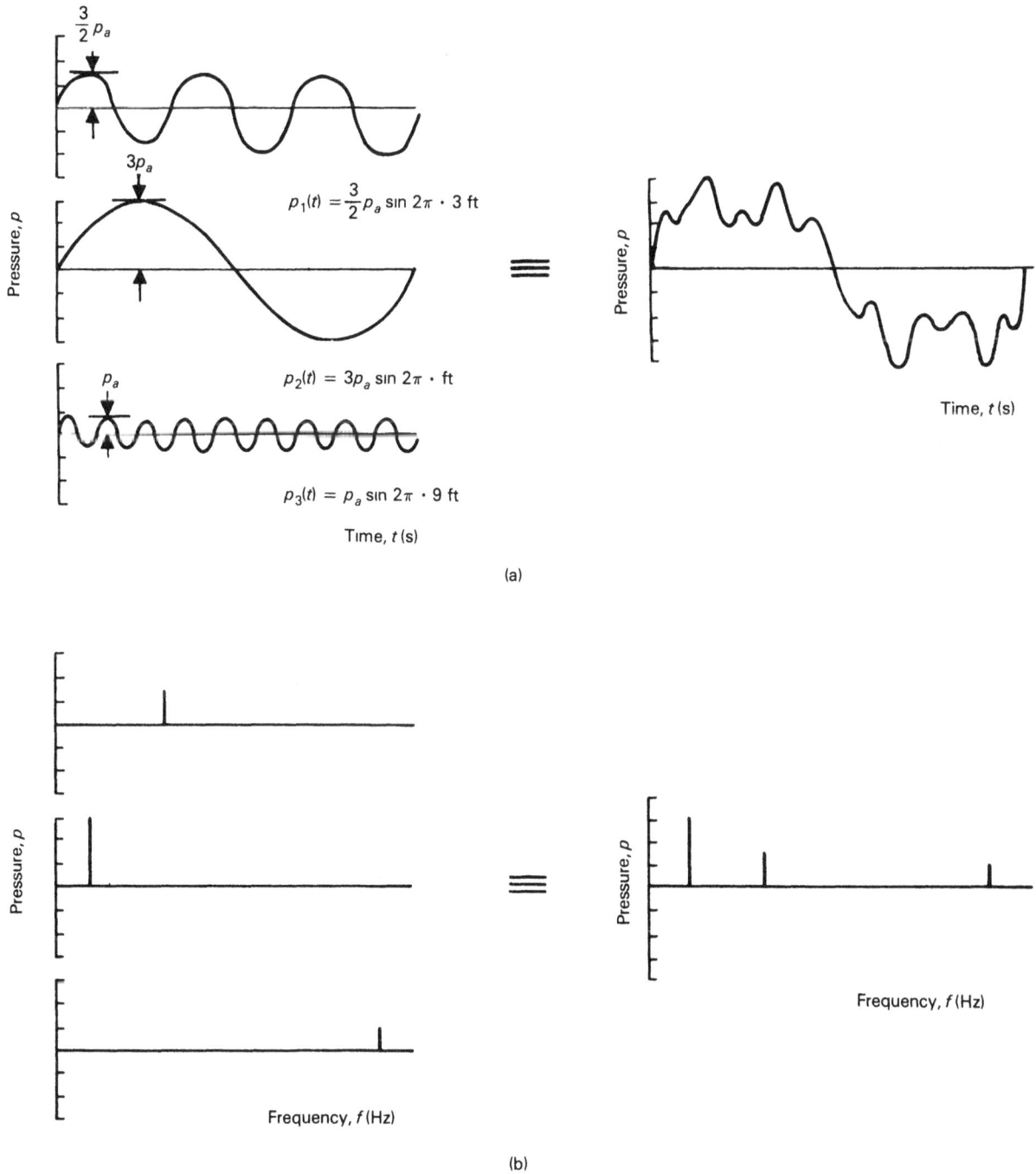

Figure 2.4 – Complex periodic waves. (a) Formation from simple pure tone components of a complex wave. (b) Corresponding frequency spectra. p_a is the ordinate scale division of sound pressure amplitude.

5

of pure tones spaced apart only infinitesimally and with differing amplitudes. The spectrum in this case is usually considered to be broadband.

Examples of periodic sources and broadband sources are, respectively, rotating machinery such as turbines, fans, etc., and aerodynamic noise associated with jets or flow past a solid surface. In fact, the two types often occur together with a series of pure tone peaks above a broadband spectrum. An example is where a large machine with several random sources has a strong periodic mode of operation. For more specific detail on the frequency spectra and appropriate analysis required for different kinds of sources see section 3.3.3.

2.1.3 Wavelength

A wavelength is defined as the distance between analogous points on successive waves of a pure tone signal; it is most easily assessed by measuring the distance between adjacent maxima or minima on the pressure spatial distribution (see figure 2.3a, above). Wavelength is denoted by the symbol λ and is numerically equal to the ratio of the speed of sound in a medium to the frequency:

$$\lambda = c/f \quad \text{m}$$

where
  c is the speed of sound, m/s
  f is the frequency, Hz.

2.1.4 Speed

For very high amplitude waves, which occur either when the ambient speed of sound is exceeded (see section 7.1.1) or with strong source radiation as in certain severe impact situations, shock formation occurs, and the waveform is distorted into nonlinear behavior. Essentially, this means that sounds of different frequencies propagate at different speeds through the transmitting medium. However, for most commonplace situations in industry this will not be so, and the speed of sound can be regarded as being the same for all frequencies.

For propagation in a gas, the general expression for the velocity of sound waves, c, is given by

$$c = \sqrt{\gamma r T} \quad \text{m/s}$$

where
  γ is the ratio of specific heat of gas at constant pressure to that at constant volume

r is the constant dependent on gas involved and equal to the ratio of the gas constant and molecular weight of the gas, J/kg·K
T is the absolute temperature, K.

The velocity is directly proportional to the square root of the absolute temperature, the other terms being constant. Pressure changes and non-uniformity of gas composition are of much less significance in affecting the velocity.[1] For air at atmospheric pressure, the relationship simplifies to

$$c = 20.05\sqrt{T} \quad \text{m/s}$$

More complex relationships exist for liquids and solids[1]:

Liquids:
$$c = \sqrt{\frac{\gamma B_{bt}}{\varrho_0}}$$

Solids:
(large cross section)
$$c = \sqrt{\frac{B_b + \frac{4}{3}B_s}{\varrho_0}}$$

Solids:
(small cross section)
$$c = \sqrt{\frac{B_y}{\varrho_0}}$$

where
  B_{bt} is the isothermal bulk modulus, N/m^2
  ϱ_0 is the density, kg/m^3
  B_b is the bulk modulus, N/m^2
  B_s is the shear modulus, N/m^2
  B_y is the stiffness (Young's) modulus, N/m^2.

These equations apply for longitudinal waves only. For transverse waves in solid bars and plates, where flexural motion involves the displacement of the structural elements perpendicular to the direction of propagation, the velocity is also dependent on the frequency of the waves and is therefore not simply described. This gives rise to the phenomenon of coincidence, important for its deleterious effect on sound transmission properties (see section 5.2.1.2). Note that all the above equations indicate an inversely proportional relationship of velocity to density. Velocity often increases with density because for many materials it is accompanied by a concomitant increase in elasticity, represented by the various moduli concerned.

Table 2.1 gives examples of sound velocity in several materials and transmission media frequently encountered in noise control. All values are quoted for 20° C unless otherwise noted.

Table 2.1 – Sound velocities for some materials common in noise control.

Material	Additional factors	Sound velocity (m/s)
Air	—	343
Steam	100° C	405
Water	—	1408
Aluminum	Bar	5150
Concrete	Bulk	3100
Steel	Bar	5050
Steel	Bulk	6100
Lead	Bulk	2050
Glass	Bar	5200
Wood (soft)	Bulk	3500
Wood (hard)	Bulk	4000

2.1.5 Spherical Wave Propagation

Consider a balloon expanding and contracting in phase at all points on its surface about some equilibrium size. The speed of propagation of sound waves away from the surface is the same as for a plane wave in a tube. In the latter situation the sound pressure along the tube is independent of distance (assuming no frictional losses), whereas in the case of a balloon it is dependent on the radial distance, r, of the wavefront from the center of the balloon. This is because the area of the wavefront increases with r and so, as shown in section 2.2.3, the intensity must decrease (see also figure 3.10).

2.2 PHYSICAL QUANTITIES AND UNITS

2.2.1 Root Mean Square (RMS) Sound Pressure

For sounds which are not pure tone, it is not sufficient to characterize the associated pressure fluctuations with a simple measurement of the maximum amplitude attained. This is because the waveform is not simple, and maxima and minima are not necessarily of equal amplitude. Since most common sounds consist of a series of rapid positive and negative pressure displacements about the equilibrium pressure value, straightforward computation of the arithmetic mean value would lie very close to zero.

Ideally, a measurement of sound pressure should take account of the magnitude of excursions from the equilibrium value on both sides, especially for irregular sounds. This measurement should also represent some kind of time-average

of the sound pressure. It is achieved by squaring the instantaneous pressure values over a specified period of time and then taking the square root.[2] The process of squaring converts all the negative pressure values into positive quantities, which can then be summed with all the squared positive values. An additional advantage is that energy is related directly to the square of pressure. The process of finding the root mean square pressure p_{rms} is represented by

$$p_{rms} = \left[\frac{1}{T} \int_0^T p(t)^2 \, dt \right]^{1/2} \ \text{N/m}^2$$

where

$p(t)$ is the instantaneous sound pressure
$\int dt$ is the integration over infinitesimal intervals of time, dt
T is the time period concerned

For simple sinusoidal waveforms (see figure 2.3), the RMS pressure is related to the maximum peak pressure p_a by

$$p_{rms} = \frac{p_a}{\sqrt{2}} \ \text{N/m}^2$$

It is important to note that an RMS measure retains a constant value regardless of the relative phases of constituent pure tones in a complex wave. All future references in the text to pressure will implicitly refer to RMS sound pressure unless otherwise stated.

2.2.2 Sound Intensity

Intensity is defined as the average rate of flow of energy through a unit area normal to the direction of propagation; it is normally measured in watts per square meter, W/m². This quantity is often used in acoustics, since many problems are associated with identifying sound or vibrational energy transfer between component spaces and materials within a system. The principle of energy conservation allows the paths of transmission and dissipation to be tracked. For free plane waves, the average intensity measured is equal to the time-averaged product of pressure, p, and particle velocity, U, as measured in the direction of interest:

$$I = \overline{pU} = \frac{p_{rms}^2}{\varrho c} \ \text{W/m}^2$$

2.2.3 Sound Power

The rate of flow of energy that is transmitted through a medium surrounding a source generating sound or vibrational energy is termed power. When there are no losses in the medium itself, then any surface which completely encloses the source must have all the power passing through it. An analogy with heat may be drawn here. Although a radiator or flame may have a certain thermal energy measured in watts which characterizes that particular heat source, the perception of the effects of energy radiating from that source is in terms of temperature, which is partly a function of the surrounding medium. So it is with sound, except that its effects are characterized in terms of pressure or intensity.

As the hypothetical surface enclosing a sound power source is made larger, less power per unit area or intensity passes through any portion of the surface (see figure 3.10). The total sound power measured in watts (W) is equal to the product of the intensity and the area of the surface; this is assessed by summation of these products over optimal-sized elements of the surface chosen. (Refer to sections 3.4.2 and 3.4.4 for a discussion of the approach used for directive sources.) Thus

$$W = \int_S I_S \, dS \quad \text{W}$$

where

S is the area of enclosing surface, m^2
I_S is the intensity through each elemental area dS, W/m^2.

With spherical waves in free space, the area of any enclosing sphere is $4\pi r^2$, and hence intensity is inversely proportional to the distance from the source, r, according to

$$W = I \cdot 4\pi r^2 \quad \text{W} \qquad \text{(for omnidirectional sources only)}$$

In the far field of a spherically radiating source (see figure 3.11), the relationship of sound intensity and pressure is the same as that given in section 2.2.2.

2.2.4 Energy Density

Energy density is significant where the sound field is composed mainly of standing waves (see section 2.3.3), rather than free progressive waves, whose phase relationship with reference to any particular point is always changing. This can occur in ducts and at low frequencies in enclosed spaces. The total energy density consists of two parts: one associated with the kinetic energy of particles contained within a unit volume, and another associated with the potential energy of the gas within that volume. The expression for energy density at a particular point is

$$D(x,t) = \varrho U^2 + \frac{p^2}{\varrho c^2} \quad \text{W·s/m}^3$$

where

U is the rms particle velocity, m/s
p is the rms acoustic pressure, N/m^2
ϱc is the characteristic impedance of air ($= 415$ N·s/m^3)
c is the velocity of sound in air ($= 343$ m/s).

When the pressure and velocity are space-averaged this reduces to

$$D(t) = \frac{\bar{p}^2}{\varrho c^2} \quad \text{W·s/m}^3 \text{ (or J/m}^3)$$

where

$\bar{p}$ is the space-averaged rms acoustic pressure, N/m^2.

2.2.5 Acoustic Impedances

Acoustic impedances are usually expressed as the complex ratio of sound pressure to particle velocity at a particular location, usually the surface of an acoustic absorber. Thus

$$Z = \frac{p}{U} \quad \text{N·s/m}^3 \text{ (mks rayls)}$$

where

p is the complex acoustic pressure, N/m^2
U is the complex particle velocity, m/s
Z is the specific acoustic impedance, N·s/m^3.

Complex impedances may also be represented in the form

$$Z = R + iX$$

where

    R is the resistance (or real) component
    X is the reactance (or imaginary) component
    i is the $\sqrt{1}$.

Impedances are of particular importance because they relate directly to the ease with which a sound wave or vibration may be transferred between two media. The resistance represents the relative magnitude of an acoustic wave which passes through a medium interface, whereas the reactance takes into account any phase changes which occur. When dealing with porous acoustic materials, the above definition is refined somewhat because of the variety of angles which the alternating fluid particle velocity may assume.[3] The normal specific acoustic impedance, Z_n (see section 5.1.1), is defined using the normal component (perpendicular to air-material boundary) of the particle velocity, U_n; thus

$$Z_n = \frac{p}{U_n} \text{ N} \cdot \text{s/m}^3$$

For free progressive plane waves, or spherical waves at a large distance from the source, the relationship between sound pressure and particle velocity is expressed as

$$\varrho c = \frac{p}{U} \text{ N} \cdot \text{s/m}^3$$

where

    ϱ is the density of air, kg/m^3
    c is the speed of sound propagation in air, m/s.

The term ϱc is sometimes known as the characteristic impedance; it occurs frequently in acoustic equations. Under normal atmospheric temperature and pressure, 20° C and 1.013×10^5 N/m^2, its value is 415 N$\cdot$s/m^3.

2.3 ADDITIONAL FEATURES ASSOCIATED WITH THE WAVE NATURE OF SOUND

2.3.1 Diffraction

Sound waves radiating from a source are reflected or diffracted by objects in their path in much the same way as waves on the surface of water. Diffraction is the process by which sound

waves experience a change in their direction of propagation after encountering an obstacle; it is equivalent to a bending effect. For most general noise control purposes, if one or more major dimensions of an intervening object are of the same order of magnitude or larger than the wavelengths associated with the frequencies of interest, then the object will act as a barrier to the sound. Otherwise, diffraction will occur around the edges of the object as illustrated in figure 2.5. The effect typically means that high-frequency waves tend to be much more directional in character; their path of propagation can often be successfully traced using a geometric ray approximation. Diffraction is particularly important when examining the influence of microphones and barriers on sound fields (see sections 4.1.4 and 6.2.1.1, respectively).

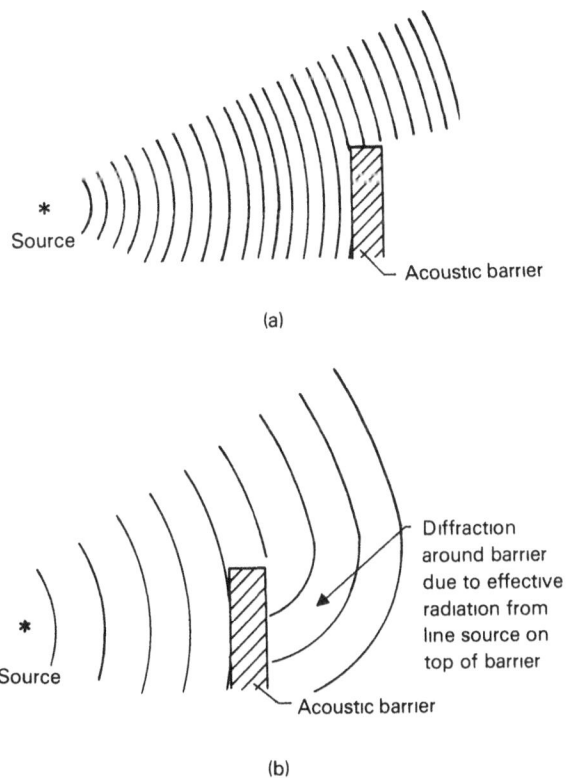

Figure 2.5 – Diffraction effects of a solid obstacle. (a) High frequency. (b) Low frequency.

2.3.2 Superposition

The principle of superposition, which applies to waveforms with amplitudes typical of industrial situations, states that the net result of adding two waves is to sum their effects arithmetically. In figure 2.6 this is illustrated for two sets of waves

9

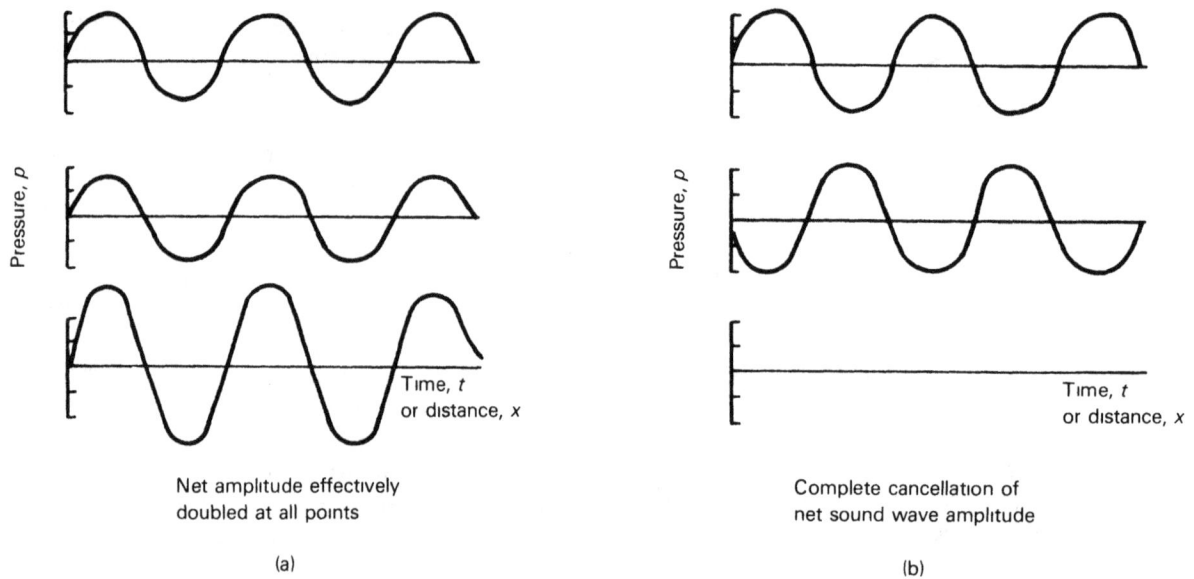

Figure 2.6 – Addition of two identical sine waves, with (a) 0° and (b) 180° phase differences.

which are identical in all respects except that their phase differences are respectively 0° and 180°.

This result flows from the addition of n linear equations:

$$p(x,t) = p_1(x,t) + p_2(x,t) + \cdots + p_n(x,t)$$

2.3.3 Standing Waves or Resonances

When a free progressive wave encounters a sharp change in impedance such as at an air-hardwall boundary, and if the angle of incidence is normal, then the reflected wave travels back along the original path of propagation with a 360° change in phase. If there is another boundary parallel to the first one, then the reflection process occurs again, and so on ad infinitum, until all the acoustic energy has dissipated by absorption at the boundaries and by propagation losses in the intervening medium. Standing waves or resonances occur when the distance between the boundaries coincides with an integer multiple of the half-wavelength; they are more likely to appear when the original complex waveform has strong discrete components.

By superposition, two waves 360° out of phase add constructively to produce a waveform greater in amplitude than the original constituents. In this unique case, however, because a complete reversal of direction has also taken place, fixed positions are created of zero net pressure amplitude (nodes) and of maximum pressure amplitude

(antinodes). At intermediate points, the amplitude varies sinusoidally with a fluctuating range described by the envelope in figure 2.7 at a rate equal to the frequency of the contributing waves. Because of the fixed pattern of amplitudes in space, the resultant waveform is called a standing wave (figure 2.7).

Standing waves most commonly occur in any regular enclosure which possesses at least one set of parallel walls. The criterion for the frequencies at which they may become established is related to the separation distance between parallel surfaces by

$$f_n = \frac{nc}{2d} \quad \text{Hz}$$

where
    n is the $1, 2, 3 \ldots$
    c is the velocity of sound propagation, m/s
    d is the separation distance, m
    f_n is the nth harmonic frequency, Hz.

Note that the term n is included because harmonics of standing waves may occur at integer multiples of the fundamental.

This phenomenon is of particular importance in the transmission of vibrational energy through plates and bars, as well as in airborne sound transmission through limited spaces of regular geometry. In both situations, reflection relatively free of energy loss at the metal or room boundaries may

10

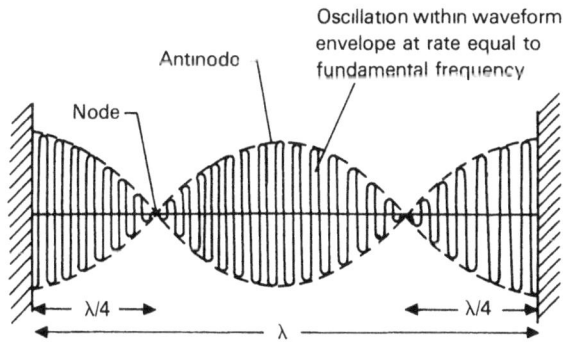

Figure 2.7 - Standing wave between two parallel surfaces one wavelength apart.

occur, with a phase change of approximately 360°. In practice, therefore, particular structures and spaces may tend to amplify acoustic energy at those frequencies where standing waves occur. Therefore, when designing noise control procedures, for example, enclosures, particular attention must be paid to preventing standing wave formation at frequencies given by the above formula, to prevent undue degradation in predicted attenuation performance. An alternative description used for resonance in solid materials is normal modes of vibration.

2.4 DECIBELS

The parameters typically used to describe sound have very wide ranges. Frequencies may change by an order of 10^4, sound power or intensity by 10^{16}, and sound pressure by 10^8. To simplify the graphical and numerical expression of these quantities, they are represented by means of a logarithmic transformation. In the case of sound power, intensity, and pressure, this is achieved using the decibel scale.

Essentially, decibels are calculated by taking the logarithm of the ratio of the value of the parameter of interest to a fixed reference value of that parameter, and multiplying it by 10. The advantages of this are threefold: it avoids large exponents in representing the quantities; the human ear judges sounds according to the ratio of intensities, so that the subjective effect is reasonably approximated; and electrical engineers use decibels in evaluating electrical power and voltage, so that conversion of quantities and units at electroacoustic transducers is simplified. By convention, parameters are taken as "levels" when their units are expressed in decibels.

An acoustical source having a sound power of W watts is therefore said to have a sound power level PWL given by

$$PWL = 10 \log \frac{W}{W_{ref}} \quad dB$$

where

W_{ref} is the reference power value normally taken as 10^{-12} W, according to international convention.

For example, if $W = 1.3 \times 10^{-5}$ W, then
$$PWL = 10 \log (1.3 \times 10^{-5}/10^{-12})$$
$$= 10 \log 1.3 \times 10^7$$
$$= 10 \times 7.11 = 71.1 \text{ dB re } 10^{-12} \text{ W}$$

Because we have taken a ratio, decibels are expressed in dimensionless units. It is important, however, always to state clearly the associated reference to avoid confusion with possible alternatives. Sound intensity level IL is defined in a way identical to sound power level using a reference intensity of 10^{-12} W/m². Thus

$$IL = 10 \log I/I_{ref} \quad dB$$

Sound pressure level, SPL, is defined by the relation

$$SPL = 10 \log \left(\frac{p}{p_{ref}}\right)^2$$

where

p_{ref} is the reference pressure value, taken as 2×10^{-5} N/m²

or

$$SPL = 20 \log \frac{p}{p_{ref}}$$

Since most noise control problems involve a calculation of airborne sound propagation, the reference pressure, p_{ref}, is optimized so that sound pressure level and sound intensity level are numerically equal for measurements made in the atmosphere. We have already presented an equation relating sound intensity to sound pressure under certain circumstances (see section 2.2.2):

$$I = \frac{p^2}{\varrho c} \quad W/m^2$$

11

and,

$$10 \log I/I_{\text{ref}} = 10 \log \left(\frac{p}{p_{\text{ref}}}\right)^2$$

Since the reference intensity value equals 10^{-12} W/m², therefore

$$(p_{\text{ref}})^2 = 10^{-12} \times \varrho c$$
$$= 10^{-12} \times 415$$
$$(\varrho c = 415 \text{ N·s/m}^3; \text{ see section 2.2.5})$$

Hence

$$p_{\text{ref}} = 2.04 \times 10^{-5} \quad \text{N/m}^2$$

To standardize, this is rounded to give

$$p_{\text{ref}} = 2 \times 10^{-5} \quad \text{N/m}^2$$

Note that this unique equivalence between sound pressure level and sound intensity level obtains only for airborne sound; in other fluids or solids different reference values are used. For techniques of combining quantities expressed in decibel values refer to section 3.2. The relationship of sound pressure and sound intensity to decibels for airborne propagation is shown in table 2.2.

Figure 2.8 illustrates sound power levels of some typical noise sources.

Table 2.2 – Relationship between decibels, sound intensity, and sound pressure for atmospheric propagation.

Sound intensity, I (W/m²)	Sound pressure, p (N/m²)	Il re 10^{-12} W/m² (dB) or SPL re 2×10^{-5} N/m² (dB)
10^4	2×10^3	160
10^2	2×10^2	140
10^0	2×10	120
10^{-2}	2×10^0	100
10^{-4}	2×10^{-1}	80
10^{-6}	2×10^{-2}	60
10^{-8}	2×10^{-3}	40
10^{-10}	2×10^{-4}	20

Sound power level PWL re 10^{-12} W (dB)

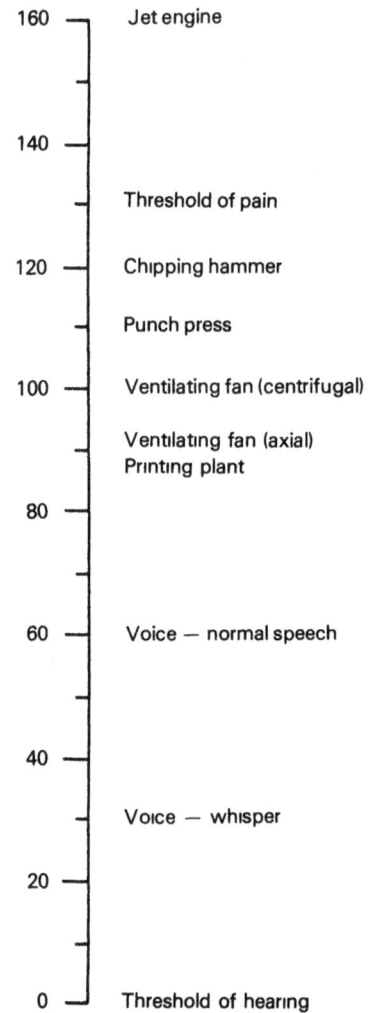

Figure 2.8 – Sound powers for a variety of sources.

REFERENCES

1. Kinsler, L. E., and A. R. Frey, *Fundamentals of Acoustics*, 2d ed., John Wiley & Sons, 1962.
2. Maling, George C., Jr., and Williams W. Lang, Theory into Practice: A Physicist's Helpful View of Noise Phenomena, Noise Con. Eng. vol. 1, p. 46, 1973.
3. Beranek, Leo L., *Noise and Vibration Control*, McGraw-Hill Book Co., 1971.

CHAPTER 3

Measurement Techniques

Sound has temporal, spectral, amplitude, and spatial characteristics; because the process by which it interacts with human physiology and behavior is by no means well specified, different indices and scales exist. This chapter initially treats in some detail the most important parameters and measurement approaches used for general-purpose industrial noise control. Section 3.4 contains a full discussion of alternative means of identifying noise sources, and section 3.5 presents the overall approach for a noise survey. These sections are precursors to chapter 6, which examines general noise control procedures that are applied after the problem is defined.

3.1 ACOUSTIC PARAMETERS OF SPECIAL RELEVANCE

3.1.1 Loudness Evaluation

It has been indicated in section 2.0 that loudness is an important subjective perceptual characteristic of interest. There are several approaches to quantifying loudness; usually the results are averaged over a number of people, since individual responses can vary considerably.

Figure 3.1 shows equal loudness contours, which represent the variation in actual sound pressure level with frequency necessary to produce an equivalent sensation of judged loudness by the listener. These curves are based on pure tone stimuli and are characterized by assigning them a phon number, which corresponds to their sound pressure level at 1 kHz. Thus, for example, a sound of 103 dB at 100 Hz represents a loudness value of 100 phons.

Several interesting features appear in these curves. The dip around 3–4 kHz corresponds to an increase in sensitivity resulting mainly from a resonance effect in the ear canal. Also, the spacing between adjacent curves is roughly constant, at

8–10 dB, across the frequency range of most interest in noise control, 300 Hz to 4 kHz. This allows the generalization that a reduction of approximately 10 dB is needed to reduce loudness by one-half, since a change of 10 phons represents a doubling or halving of loudness. In typical working conditions, however, an observer is rarely able to detect a change of less than 3 dB in levels. Especially prominent is the decrease in sensitivity at all levels for sound at the lower frequencies. This means, for example, that a sound of 80 dB at 30 Hz produces the same sensation of loudness as a tone of 42 dB at 4 kHz.

A more analytical approach has been taken by both Zwicker and Stevens,[2] whose models are predicted on the observation that the ear appears to act as a set of filters normally around one-third octave in bandwidth (see section 3.3.1 for definition). Their basic contention is that subjective loudness should be obtainable by summing the sound pressure in these bands after correcting for the sensitivity of the ear. This varies with frequency and level. The Zwicker and Stevens MK. VI tech-

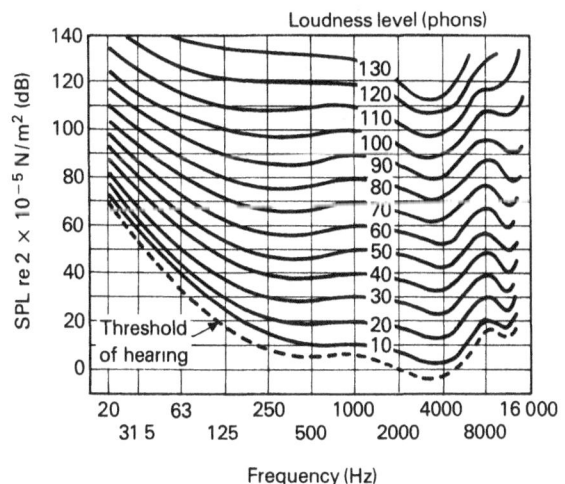

Figure 3.1 – Equal loudness contours (reference 1).

niques have found widespread use and have been adopted as international standards.[2] More recently, Howes has proposed a model[3] for calculating the loudness of sounds that are reasonably steady over a period of at least a second or so. This model uses power summation over the critical bands in the ear and the concept of frequency interaction. Computation techniques for Zwicker and Stevens' phons may be compared by examining reference 4, and Howes has developed a computer program[5] for general dissemination which requires as input only the sound spectrum and mode of listening.

Apart from the common-sense assumption that a loud sound tends to be more damaging in regard to hearing loss, it is also reasonable to suppose that, other things being equal, a louder sound tends to be more disturbing. Various researchers have tried to characterize sound in terms of annoyance or noisiness ratings, which presumably represent a more comprehensive subjective evaluation than loudness per se. Reference 6 contains an early description of such a procedure as well as a comparison with loudness calculation techniques.

3.1.2 Weighting Scales

In order to assess human response, it would be convenient to use a single numerical criterion which can be easily measured. Initially, it was proposed that this be achieved by inverting the 40-, 70-, and 90-phon curves in figure 3.1 and providing electronic filter characteristics in sound level meters and other devices which closely approximate these shapes. This permitted integration of sound pressure over the range of interest, providing a measure of subjective response to low, medium, and high level sounds, respectively. The corresponding filter networks are designated A (40 phon), B (70 phon), and C (90 phon). The relative responses of these are illustrated in figure 3.2

Subsequently, it was found in practice that, although anomalous, the use of the A-weighting scale gives results which correlate best with human response in regard to noise-induced deafness, the acceptability of noise from transportation and other machines, and communications interference. Measurements on the various scales are denoted by the symbols dB(A), dB(C), etc.; of course, the reference value should be specified where necessary. Table 3.1 shows the approximate numerical value of the two most popular weightings relative to linear response, applied in octave bands (see section 3.3.1).

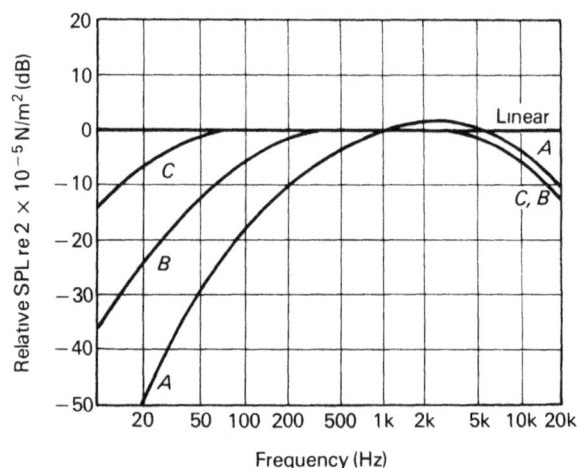

Figure 3.2 – Frequency response of weighting scales. The straight line through 0 dB represents the unweighted response normally used when assessing linear overall sound pressure level, OASPL.

Table 3.1 – A- and C-weighting corrections for octave band data.

Octave band center frequency (Hz)	Relative A-weighted response (dB)	Relative C-weighted response (dB)
31.5	− 39.5	− 3.0
63	− 26.0	− 1.0
125	− 16.0	0
250	− 8.5	0
500	− 3.0	0
1 000	0	0
2 000	+ 1.0	0
4 000	+ 1.0	− 1.0
8 000	− 1.0	− 3.0
16 000	− 6.5	− 8.5

3.1.3 Other Indices and Scales

As has already been implied in section 3.1.1, sounds may be identified as noise resulting from a variety of factors. Several techniques for assessing the effects of noise in specific situations have evolved, and it is worthwhile to mention a few that can be relevant to industrial noise problems. A fuller description of these and others, including applications and calculation procedures, may be found in reference 4.

A technique widely used in the United States for assessing the relative intrusiveness of background noise is based on a set of contours known as Noise Criteria curves (see figure 3.3). These are similar to equal loudness curves and have been

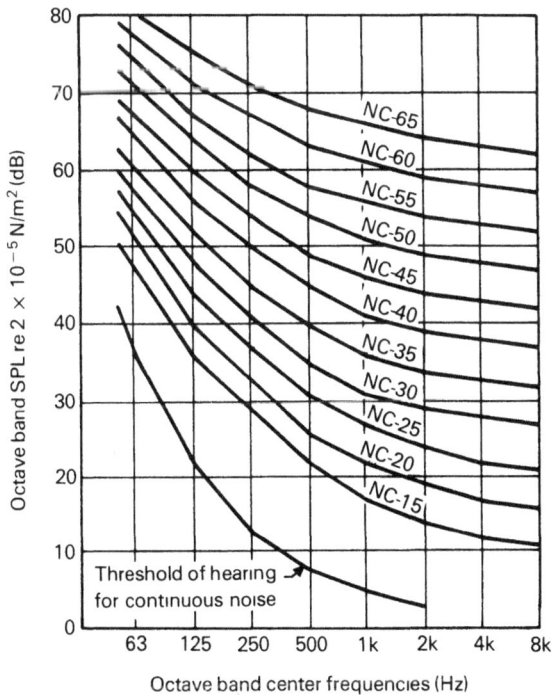

Figure 3.3 - Noise criteria curves (NC).

developed after considerable practical experience. They have essentially been optimized to provide an assessment of the effects of noise at particular frequencies on speech communications, with some regard to the loudness of the sound as well.

A Noise Criteria (NC) rating is found by plotting the actual octave band sound pressure levels and identifying the highest NC curve which penetrates the sound spectrum from figure 3.3. Certain criteria, shown in table 3.2, have been recommended for particular environments.[7]

The equivalent continuous sound level, L_{eq}, is increasingly used in a wide variety of situations.

Table 3.2 - Recommended noise criteria values for several work environments.

Work environment	Probable acceptable NC levels
Factories and workshops (no hearing damage risk)	60–75
Factories and workshops (communication required)	50–60
Kitchens and light workshops	45–55
Mechanized offices	40–50
General offices	35–45
Private offices	30–40
Cafeterias and canteens	35–45

Not the least of the reasons for this is the belief that the amount of sound energy received at the eardrum is the most critical cause of noise-induced hearing loss. L_{eq} takes the value of the constant sound pressure level which contains the same sound energy as the actual time-varying level of interest. It is always calculated with respect to a particular time interval, generally an hour or an 8-hour period; normally, the sound level is A-weighted. With digital and microprocessor technology, a variety of portable instruments can accurately compute L_{eq} (see section 4.4.1).

The mathematical expression for the computation of L_{eq} over a specified time interval T is

$$L_{eq}(T) = 10 \log \frac{1}{T} \int_0^T \frac{p^2}{p_{ref}^2} \, dt \quad \text{dB}(A)$$

where $\int_0^T dt$ represents the integration or summation of squared instantaneous sound pressure values over the time interval T

p is the A-weighted instantaneous sound pressure

p_{ref} is the reference sound pressure.

Community noise can often be a problem when external noise levels are high. Consideration should be given to this especially when planning a new industrial development or implementing noise control technology in a situation where community aggrievance has already been established. A very concise introduction to the subject may be found in reference 8.

One of the indices most frequently used in this connection[9] includes the day–night level, L_{dn}, which represents a long-term average 24-hour L_{eq}, with sound energy received between 10:00 p.m. and 7:00 a.m. being weighted by a factor of $+10$ dB. The community noise equivalent level, CNEL, performs a similar function except that a three-way subdivision between day, evening, and night is used. Statistical measures such as L_{90}, L_{50}, and L_1, illustrated in figure 4.10, are also widely used. L_{90}, or the level exceeded for 90% of the sampling time is often used as a measure of ambient noise.

3.2 LOGARITHMIC ADDITION OF DECIBELS

It is frequently necessary to combine several sound levels in noise control problems; this is done by taking antilogarithms and summing with respect to mean square sound pressure. The most important initial step is to determine whether the sources concerned radiate sound intensity coherently, because pure tones of the same frequency can interfere constructively, as described in section 2.3.2, and it is necessary to account for the phase relationship between the sources, which is fixed for coherent radiation. The appropriate equation for calculating the RMS sound pressure in this case is

$$p = \sqrt{p_1^2 + p_2^2 + 2p_1p_2 \cos(\theta_1 - \theta_2)} \quad \text{N/m}^2$$

where

p_1, p_2 is the RMS sound pressure associated with the two sources

$(\theta_1 - \theta_2)$ is the relative phase difference between sources at measuring point.

Note that for completely in-phase signals of equal amplitude this has the effect of doubling the resultant pressure. The total sound pressure level in this case would be raised 6 dB with respect to the original pressure:

$$p = \sqrt{p_1^2 + p_1^2 + 2p_1^2} \quad \text{N/m}^2$$
$$= 2p_1$$

Therefore,

$$\Delta \text{SPL} = 20 \log \frac{2p_1}{p_1} = 6 \text{ dB}$$

For most industrial situations, it is reasonable to assume that the sources radiate sound randomly with respect to each other (i.e., are incoherent), especially when only weak pure tones are present. Under these conditions, phase relationships become unimportant and the squares of the RMS pressures of the n contributing sources at the point of interest may be added linearly:

$$p = \sqrt{p_1^2 + p_2^2 + p_3^2 + \cdots + p_n^2} \quad \text{N/m}^2$$

The squares of pressures in the above equations are taken before addition because it is important to observe the principle of conservation of energy, and energy or intensity is proportional to the square of sound pressure. In terms of sound pressure levels this may be expressed as

$$\text{SPL}_T = 10 \log \left[10^{\text{SPL}_1/10} + 10^{\text{SPL}_2/10} + \right.$$
$$\left. \cdots + 10^{\text{SPL}_n/10} \right] \quad \text{dB}$$

where

SPL_T is the total sound pressure level at the measuring point.

Two levels may be most simply combined by use of figure 3.4. The numerical difference between the levels concerned is first found, and then the corresponding position on the curve identified. From this, after alignment with the relevant value on either the abscissa or ordinate, the total level is computed by straightforward addition.

Two propositions should be noted regarding figure 3.4:

1. For two levels with the same value, the numerical increase between the total and the original values is 3 dB.

2. For levels more than 10 dB apart, the contribution of the lower level is negligible.

Figure 3.4 is plotted according to the relationship

$$\text{SPL}_T = \text{SPL}_1 + 10 \log \left(10^{(\text{SPL}_2 - \text{SPL}_1)/10} + 1 \right)$$

where

SPL_T is the total level

SPL_1, SPL_2 indicates that two levels are being added

For example: If we have a series of levels 46, 56, 61, and 61 dB, figure 3.4 allows computation of the total as 64.7 dB. Thus

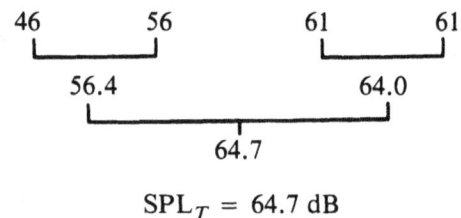

$$\text{SPL}_T = 64.7 \text{ dB}$$

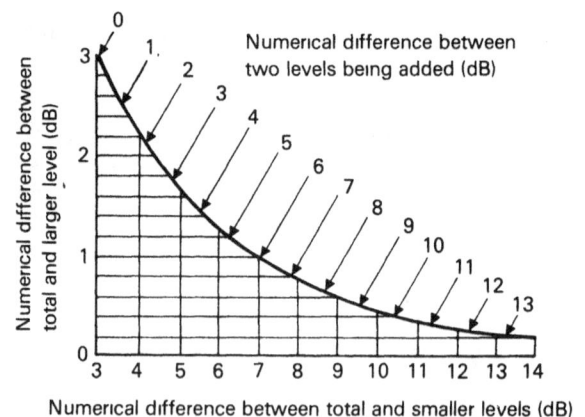

Figure 3.4 – Combination of two decibel levels.

A chart for the addition of equal levels is presented in figure 3.5. This is an exercise frequently required, especially when assessing the contributions of several identical machines to a reverberant field.

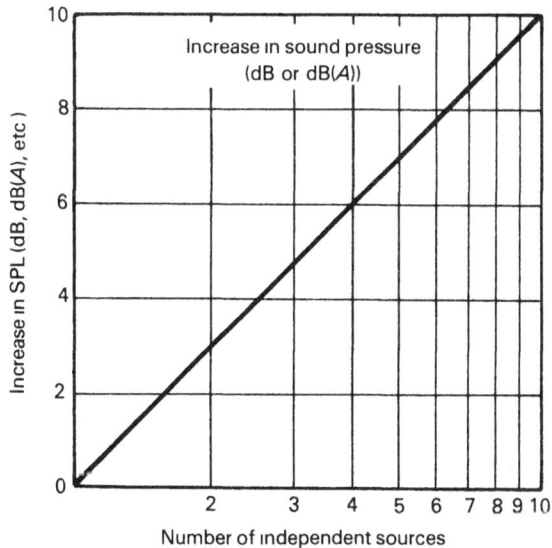

Figure 3.5 – Combination of identical levels. (From Lewis H. Bell, Fundamentals of Industrial Noise Control, 1973. *Used with the permission of Harmony Publications.)*

Average levels are computed according to the principle used above; that is, the sum of the associated energy quantities is taken, usually pressure squared, before taking the average value and converting back into decibels. Thus an average sound pressure level $\overline{SPL}$ would be computed for n separate levels according to

$$\overline{SPL} = 10 \log \frac{1}{n} \left[10^{SPL_1/10} + 10^{SPL_2/10} + \cdots + 10^{SPL_n/10} \right] \quad dB$$

where

SPL_n is the nth sound pressure level to be added.

Another computation frequently required is the addition of sound pressure levels in contiguous frequency bands to produce an overall sound pressure level. In the example below, the A-weighted overall level is calculated from linear octave band data using the numerical weightings taken from table 3.1.

	Octave band center frequency (Hz)						
	63	125	250	500	1000	2000	4000
SPL, dB	65	75	77	80	83	81	79
A-weighting, dB	−26	−16	−8.5	−3	0	+1	+1
SPL (dB(A))	39	59	68.5	77	83	82	80

69   84   84

87

Therefore total sound pressure level is given by

$$SPL_T(63\text{–}4000 \text{ Hz}) = 87 \text{ dB}(A)$$

Reference 10 provides mathematical techniques for combining decibel levels of either octave bands or constant bandwidths (see section 3.3) into an overall spectrum level.

3.3 FREQUENCY ANALYSIS

3.3.1 Description of Octave, One-Third Octave, and Narrow Bands

In order to divide the frequency spectrum conveniently into bands which can be modeled by electronic filters and analyzed independently of each other, the concept of octaves is utilized. From an acoustician's point of view, the octave is significant because it represents a doubling of frequency. Examination of table 3.3 reveals that the preferred octave and one-third octave bands in use are based on geometric mean center frequencies, f_c; thus

$$f_c = \sqrt{f_n \times f_{n+1}} \quad Hz$$

where

f_n and f_{n+1} are the lower and upper band limits, respectively.

The octave bands of most interest for general-purpose use extend from 63 to 8000 Hz; any overall sound pressure level computed from measurements made in a series of contiguous bands should properly have the range of bands quoted alongside.

One-third octave band filters are the second most common variety in widespread use; they increase the degree of resolution obtainable in the field without making the process of data collection too cumbersome (see section 3.3.2 for discussion of appropriate filter selection). Because the energy contained in each one-third octave band is proportionally less than that in the octave band which

Table 3.3 – Octave and one-third octave center frequencies and band limits (reference 11).

Octave band			One-third octave band		
Lower band limit*	Center frequency (Hz)	Upper band limit*	Lower band limit*	Center frequency (Hz)	Upper band limit*
			18	20	22
			22	25	28
23	31.5	45	28	31.5	35
			35	40	45
			45	50	56
45	63	90	56	63	71
			71	80	90
			90	100	112
90	125	180	112	125	141
			141	160	178
			178	200	225
180	250	355	225	250	282
			282	315	355
			355	400	450
355	500	710	450	500	560
			560	630	710
			710	800	890
710	1000	1400	890	1000	1120
			1120	1250	1400
			1400	1600	1780
1400	2000	2800	1780	2000	2240
			2240	2500	2800
			2800	3150	3550
2800	4000	5600	3550	4000	4470
			4470	5000	5600
			5600	6300	7080
5600	8000	11 200	7080	8000	8900
			8900	10 000	11 200
			11 200	12 500	14 130
11 200	16 000	22 400	14 130	16 000	17 780
			17 780	20 000	22 400

*Rounded off slightly.

covers that particular range, one-third octave band spectra always have values below the corresponding octave band spectra when plotted on the same graph (see figure 3.6). Reference 10 provides information on converting acoustical information from one octave set into another, and computing one-third octave band levels knowing the octave level and the decibel-per-octave slope of the spectrum.

In contrast to octave and fractional octave bands, which increase in bandwidth by a fixed percentage as the center frequency increases, narrow band filters have a constant bandwidth throughout the entire spectrum. They may be tuned to measure several different bands, typically 10, 5, 2, or 1 Hz in width. With the advent of real time ana-

lyzers suitable for deployment in the field, narrow band analysis is more practical. Nevertheless, the total spectrum bandwidth may have to be limited to provide the degree of resolution required at low frequencies particularly, since typically these instruments only have 400 bands over their frequency range (see section 4.6.2).

Sometimes it is convenient to measure a parameter known as the power spectral density, PSD, which is plotted as a function of signal power in a frequency bandwidth divided by that bandwidth, against frequency. When the PSD is computed using a sufficiently high resolution frequency analysis, then the difference in PSD functions between broadband and discrete spectra becomes sharply evident, as the former have a

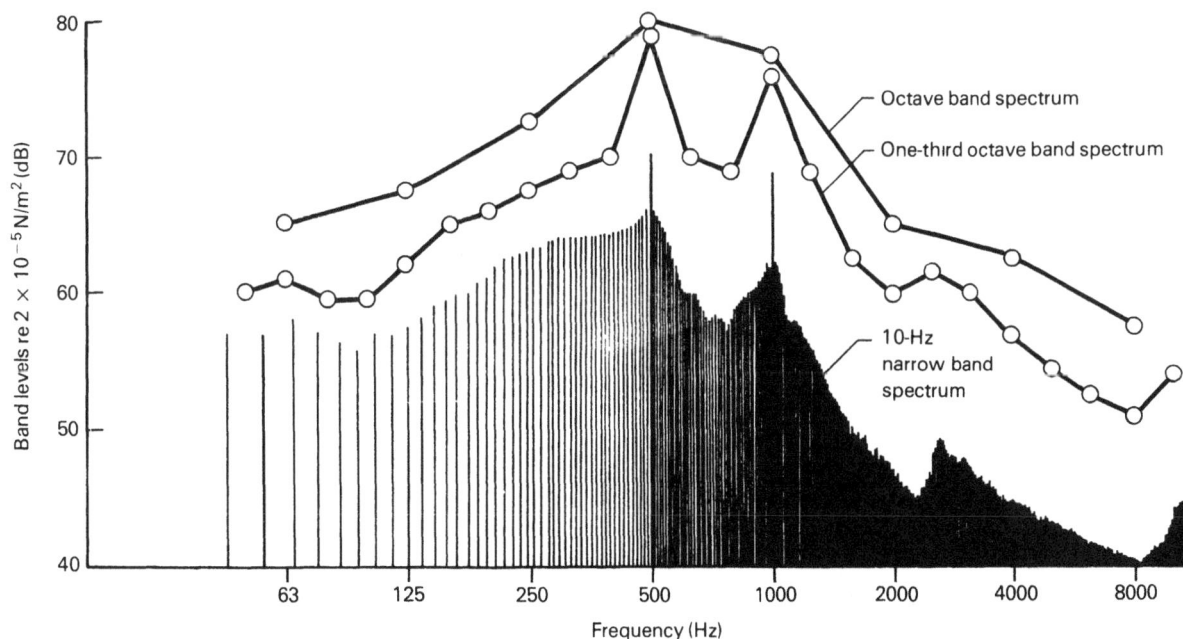

Figure 3.6 – Comparison of octave, one-third octave, and narrow band spectra for the same sound field measurement.

characteristically smooth and monotonic nature whereas the latter are very discontinuous.

There are several quick procedures for obtaining approximate spectral content. It is clear from figure 3.2 that the C-weighting curve is effectively linear over much of the frequency range of interest in general noise control. Consequently, by taking both A- and C-weighted readings and comparing them, a rough assessment can be made according to the following rules:

dB(C) > dB(A) — indicates low frequency mainly

dB(A) > dB(C) — indicates high frequency mainly

dB(A) ≃ dB(C) — indicates that it is located around the 1–2 kHz region, i.e. mid-frequency.

Many frequency filters can now be used in conjunction with weighting networks to provide a direct readout, in dB(A) for example, in specific frequency bands.

3.3.2 Description of Filtering Techniques

Various modes of filtering may be employed for either the fixed percentage bandwidth or fixed bandwidth filters described in the previous section. More detail on the instrumentation available is given in section 4.6.

The traditional mode used in the field is the serial filter, where the signal is processed by each bandwidth filter, one at a time, usually over a contiguous frequency range. In the laboratory, a similar process may be carried out using a sweep filter, where the bandwidth of the filter expands automatically as the center frequency moves continuously up through the spectrum. The latter technique involves no discrete filter bands. Either of these serial approaches is only strictly applicable when the input signal is relatively constant in level and frequency content over the period of time taken to complete the measurement.

Less frequently used is the process of feeding a signal to a set of contiguous and parallel analog filters and then, either separately or in combination, recording the output of each. This approach does not usually have the same constraints on signal uniformity as the serial process.

Finally, in digital filtering the analog signal is quantized into discrete portions and subjected to a computation process which very efficiently performs effective frequency filtering. This can be accomplished quickly and accurately. Reference 12 provides a brief introduction to the problems in digital analysis of one-third octave bands with similar accuracy across the spectrum of interest.

19

In this paper, two different sampling rates are used to achieve a higher resolution in the lower frequency bands without wasting computation time in higher frequency bands, thereby achieving a more uniform confidence level.

3.3.3 Spectral Character of Typical Industrial Sounds and Analysis Procedure

Only rarely do sounds consist of pure tones in isolation. Generally, it is necessary to conduct a spectral analysis of the acoustic or vibration parameter of interest over a series of adjoining frequency bands using a bandwidth and mode of analysis appropriate to the problem at hand. The bandwidth should be chosen to be only as narrow as necessary for the sound spectrum concerned. This section outlines the general nature of common industrial sounds and the practical methodology for conducting routine analyses.

3.3.3.1 Discrete frequency sources

Fans, compressors, gear and mechanical transmisson elements, and transformers, give rise to spectra with very prominent discrete frequency tones, as illustrated in figure 3.7a. These usually present a number of harmonic components which may even be larger in magnitude than the fundamental frequency. This may occur because the sound spectrum reaching the measuring position can depend on intervening structures (such as pressed metal panels around the machine) which radiate noise differentially according to the inherent sound transmission and damping characteristics of the complete system. This kind of spectrum often emanates from rotating equipment and it is possible to predict the harmonic frequencies, f_n, most likely to dominate according to the relation

$$f_n = nNK \quad \text{Hz}$$

where

N is the shaft rotational speed measured in revolutions per second

K is the number of blades, teeth, etc., attached to the rotating component

n is the integer corresponding to the harmonic frequencies, i.e., 1, 2, . . .

Because the frequencies of interest are very distinct, it is usual to conduct initially at least one-third octave band analysis followed by further refinement of the frequency resolution as required.

3.3.3.2 Broadband frequency sources

Broadband frequency sources are usually associated with high velocity gases or fluids venting to atmosphere. They induce a shearing action in the structure of the ambient air, and the resultant turbulence generates random pressure fluctuations. The spectrum is continuous, lacking any particular discrete quality, and with no fixed relationship between either amplitude or phase components. This is not to say, however, that the sound may not be characterized easily with reference to its frequency distribution. For example, in heating ducts the nature of sound at some distance from the air handling equipment is mainly low frequency; it is audible as a rumble. By contrast, high speed exhaust vents generate a spectrum which is predominantly high frequency in character and is best described as a kind of hiss. An

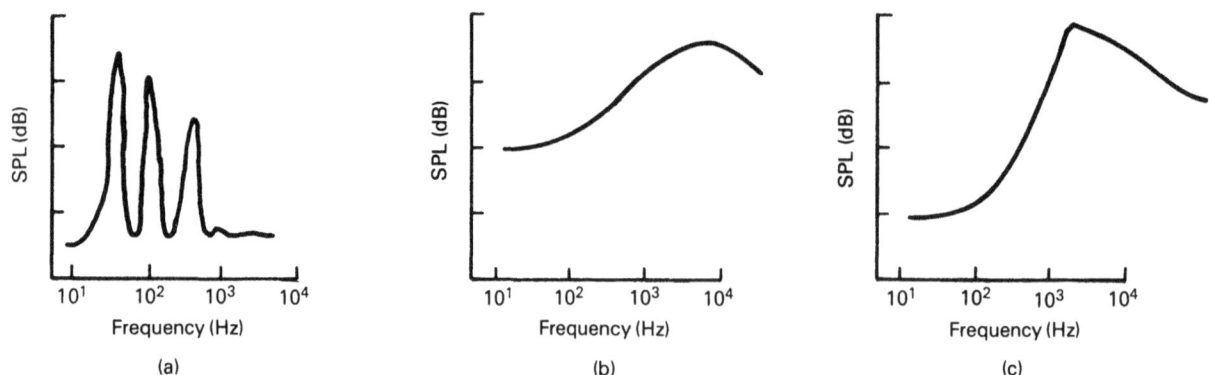

Figure 3.7 – Illustrations of various kinds of frequency spectra. (a) Discrete. (b) Broadband. (c) Impact.

20

example of this is shown in figure 3.7b. Octave band analysis will normally suffice when conducting analyses of this type of noise.

3.3.3.3 Impact sources

Impact sources include sources such as punch presses, hammering operations, and very short duration venting operations to atmosphere. The character of these may be regarded as aperiodic; consequently they have a continuous spectrum (see section 2.1.2). The distinguishing features are determined primarily by the duration of impact and the elasticity and damping properties of the materials. Basically, the shorter the duration and the harder the surfaces involved, the more likely it is that the resultant sound will have a sharp ring because of the high frequency content. This is illustrated in figure 3.7c. Because of the very short durations involved in impact noise, which can be of the order of milliseconds, the meter RMS characteristic may not respond fully to the maximum instantaneous amplitude. For this reason a peak hold characteristic with a very short integration time has been built into some modern sound level meters to gauge the peak levels involved (see section 4.4.1). Differences between this meter reading and the normal meter response can be up to 30 dB; the differences tend to increase as the distance to the source is shortened, especially for high frequency components.

3.4 GENERAL ANALYSIS

3.4.1 Source Identification

Noise control problems are usually examined by separate consideration of the source, propagation path, and receiver; an optimal solution based on an overall assessment of the results of this tripartite analysis is then designed (see section 6.0). Of these three considerations, correct identification of the source and its acoustical nature is the most important; otherwise, unnecessary expenditure may be incurred because of misdirected control procedures.

One of the simplest and quickest ways of accurately defining a source is to run all the noisy machines connected with a location of interest independently, i.e., one at a time. Individual noise contributions can then be compared in order to assess the most dominant or troublesome noise sources. Several sources may interact, however,

particularly in the case of large industrial plants, where several ancillary units such as hydraulic pumps and generators may be switched in concurrently.

A second procedure, which is rather lengthy but nonetheless of considerable value because of the amount of information which may be extracted, is to conduct a survey of sound pressure levels around the affected area. This is especially relevant in the case of shop floors where many sources operate simultaneously and where, perhaps for reasons of assembly-line production or constraints on the time during which machines may be shut down, it is necessary to conduct an in situ assessment. Figure 3.8 is an example of such a survey in a poultry processing plant. If the contour lines of equal sound pressure level are sufficiently well defined, it is possible to estimate at least relative sound power levels of contributing sources by calculating the area around each machine which is above a particular noise level. The sound power level is proportional to this area (see section 3.4.4). The survey is of importance, too, in making a quick assessment of the degree of compliance with OSHA noise exposure regulations (see the Appendix).

In addition, the directivity of the source may be estimated if the measurements are made in a direct field (see section 3.4.3). Occasionally, this can be done by making measurements very close to the radiating surfaces or duct openings,[13] although there are often theoretical objections to this; however, it is always wise to make follow-up accelerometer measurements of vibration on the radiating surfaces themselves. These may then be matched with the sound pressure level measurements to verify their value in measuring sound power and directivity.

NASA programs on jet noise sources and aerodynamic noise generation in wind tunnels have yielded several devices designed to determine source position, especially in turbulent flow conditions. One of these[14] uses a linear array of microphones with a digital time delay applied to each one in the series according to the acoustic frequency of interest. Unwanted noise containing reverberation, background, and wind noise components of the order of 3–10 dB is rejected in the range 63 Hz to 10 kHz. Another instrument,[15] which permits easy analysis in two dimensions without the need for large numbers of microphones in two separate array configurations, has a directional acuity such that high frequency sources

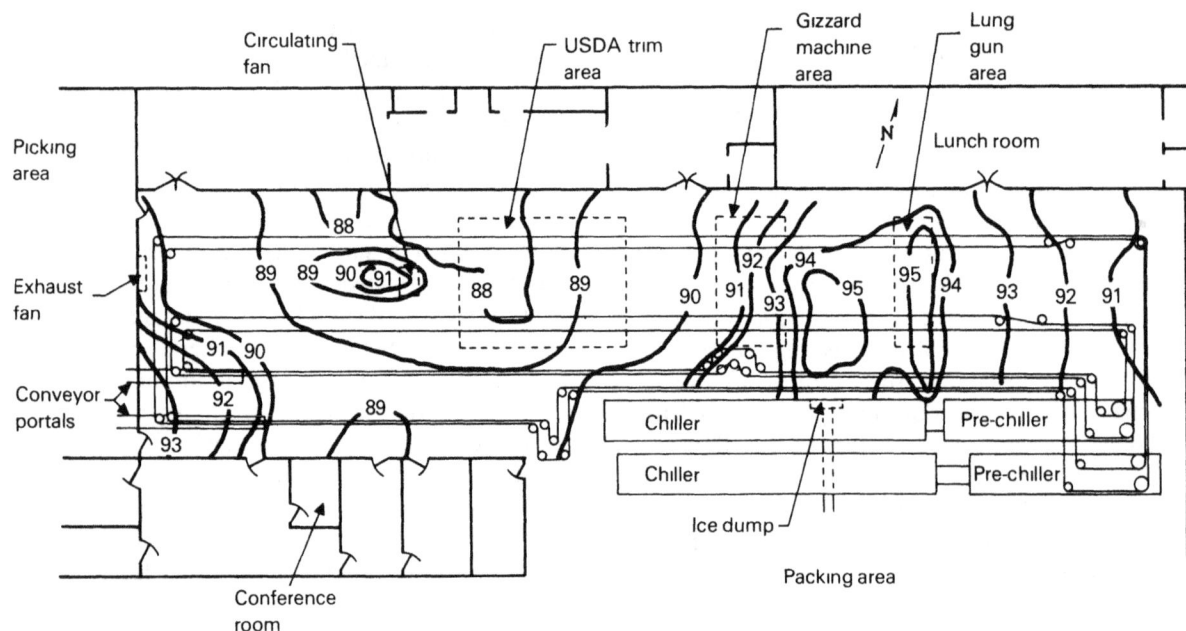

Figure 3.8 – Example of sound pressure level contour survey. All data are A-*weighted. (Taken from Cassanova, R.E., et al., Study of Poultry Plant Noise Characteristics and Potential Noise Control Techniques, Progress Report, Oct. 1978–Mar. 1979, NASA Res. Grant NSG 3228.)*

can be specified within 1 cm. This is achieved using an ellipsoidal acoustic mirror with a microphone placed at one focus; the intervening space may be filled with a gas heavier than air, to decrease the speed and hence wavelength of sound waves, so that the resolution power of the device is enhanced.[16] By mounting the mirror and microphone on an automated drive assembly, complete two-dimensional plots may be made of an area, and the frequency-analyzed output highlights significant noise sources.[17]

Another technique for determining important sources is to examine frequency plots made for measurements close to the suspected noise contributors, and to compare them with the composite level received at the position of interest. This approach is of particular value when prominent discrete frquencies are present, although narrow band analysis may be necessary to discriminate and match exact values.

More recently, many data processing techniques have been developed to extract maximum information from signals and to discard extraneous components. These can be broadly classified into either coherence or cross-correlation techniques, and their value is increasing with the increasing availability of sophisticated instrumentation for direct print-out of the parameters of

interest. An excellent account of the development of these and other techniques, as well as their application to industrial noise, is given in reference 13. Coherence and cross-correlation techniques are of particular value in situations where only in situ measurements can be made with all equipment running, and for pinpointing particularly troublesome areas on very large sources. Basically, microphones or vibration transducers are placed at the receiving point of interest and at the various suspected contributing sources, and the signals from each compared.

Cross-correlation is essentially a complete comparison of the amplitude of two signals at every frequency, as well as a measure of the time or phase difference between them. The common parts of the signals combine and the unrelated parts cancel out. By finding the delay time between the signals where the cross-correlation is a maximum, it is possible to specify noise sources and their relative contributions to the sound field by tracing possible transmission paths. Although reverberant conditions tend to militate against the effective use of cross-correlation, Ahtye and coworkers[18,19] have successfully used the technique to identify sources as low as 18 dB below the overall level. Furthermore, a measurement device known as a sound separation probe,[20] using similar principles, has been developed to separate out sound waves from turbulent flow pressure fluctuations in a duct.

Coherence measures the extent to which received power is dependent on a particular transmitted or input power. It can be used to assess the noise source characteristics in reverberant spaces,[13] as well as in situations where cross-correlation might normally be utilized. An example of the separation of contributions from a fan and aerodynamic turbulence in a NASA wind tunnel using coherence techniques is given in reference 21. The sources involved should be independent of each other,[22,23] and it is even possible to assess relative source contributions using a pair of closely spaced receiving transducers at some distance from any source, when it is not possible to mount them any closer because of environmental or geometrical constraints.[23]

3.4.2 Source Directivity

Directivity measures the extent to which the sound pressure level in any particular direction deviates from the level which would be present if the source radiated in all directions equally. Directivity of a source is normally measured in the far field and in the absence of any intervening obstacles or reflecting surfaces other than those normally associated with the source itself. It is represented by the directivity factor Q, which is defined as the ratio of the actual intensity at angle θ, I_θ, to the intensity at that point if the source were radiating omnidirectionally, I_s, i.e.,

$$Q_\theta = \frac{I_\theta}{I_s}$$

This is the same as taking the ratio of the associated sound pressure levels; in fact, the logarithmic equivalent of the directivity factor, known as the directivity index, DI, is defined as

$$DI = 10 \log Q_\theta \quad dB$$
$$= SPL_\theta - \overline{SPL_s}$$

where

SPL_θ is the sound pressure level at angle θ

$\overline{SPL_s}$ is the space-averaged sound pressure level at the same distance from the source (equivalent to assuming omnidirectional radiation).

Where complete spherical radiation is not possible because of the presence of nearby reflecting surfaces, figure 3.9 gives the values of Q and DI. The directivity index increases in value by 3 dB for every added reflecting plane. This is a natural corollary of the same sound energy flowing through half the surrounding surface area with the addition of each extra plane, i.e., a doubling of intensity.

It may sometimes be necessary to measure the directivity of a source simultaneously using a complete 360° array of microphones. This is often the case with random sources such as aerodynamic noise generated by fan or compressor blades, or an exhaust vent where a jet of high velocity gas causes turbulence in the surrounding medium. Normally, a spacing of transducers at 10° or 15° increments enables the source to be completely defined. Extensive work on aircraft jet engine noise in static tests has led to the development of a computer program for quick handling of one-third octave band data.[24,25] This program includes

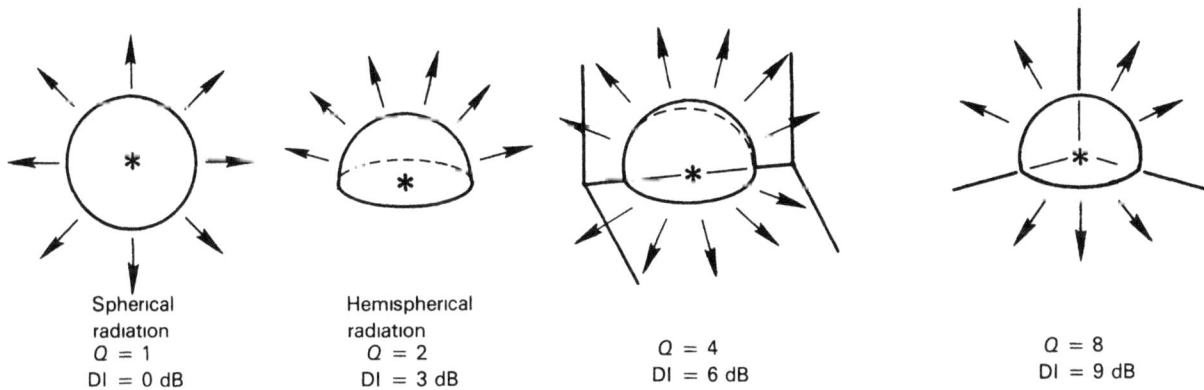

Spherical
radiation
Q = 1
DI = 0 dB

Hemispherical
radiation
Q = 2
DI = 3 dB

Q = 4
DI = 6 dB

Q = 8
DI = 9 dB

Figure 3.9 – Directivity effects resulting from reflecting planes.

standardizing and extrapolating the data as well as computing the sound power spectrum level and directivity index. Directivity effects become particularly important at higher frequencies because at low frequencies diffraction effects reduce any directive components.

3.4.3 Sound Field Characteristics

It is important when making sound pressure level measurements to consider the portion of the sound field likely to be measured. The direct or free field is that part where the sound level decays in a constant manner depending on the geometry of the source and its surroundings. For example, for a relatively small source radiating freely, i.e., with no reflecting surfaces nearby, the level SPL_r at a distance r from the source is given by

from, $\qquad W = 4\pi r_0^2 I_0 = 4\pi r^2 I \quad$ W

thus, $\qquad IL_0 - IL_r = SPL_0 - SPL_r$

$\qquad\qquad \cong 20 \log (r/r_0)$

therefore, $\quad SPL_r = SPL_0 - 20 \log r/r_0 \quad$ dB

where SPL_0 is the level at distance r_0 from the source in decibels.

This relation holds because the radiation pattern is spherical and the area through which unit intensity passes is proportional to r^2 (see section 2.2.3). The above relationship is referred to as the inverse square law and is illustrated in figures 3.10 and 3.11. The net result is a 6 dB reduction in level with each doubling of distance from the source. For line sources such as the radiating surface of a length of pipe, the radiation pattern is cylindrical and a 3 dB reduction in level with doubling of distance is found.

However, for any space confined by solid boundaries such as partitions or factory walls, the sound waves propagating freely from the reflected component contribute to an effective increase in the sound pressure level in regions relatively near the reflecting surface. The distance from the source at which this so-called reverberant field begins to dominate over the direct field is labeled r_c in figure 3.11; that figure also represents a summary of sound field characteristics for a single machine. The value of r_c depends on the acoustic power radiated and its directivity, and the absorption characteristics of the principal boundary surfaces, as well as the dimensions of the space con-

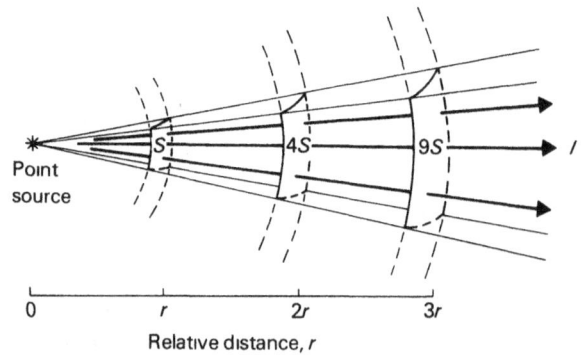

Figure 3.10 – Increase in area through which unit sound intensity passes because of spherical divergence. S, unit area; I, constant intensity through section considered.

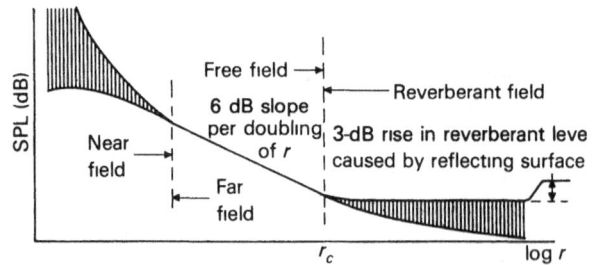

Figure 3.11 – Sound field characteristics in semi-reverberant conditions.

cerned. The reverberant field is said to be diffuse, since sound waves arrive at any point from many different directions and, except for points very close to hard surfaces, the sound level tends to be constant throughout the field.

The distinction between near field and far field is somewhat more subtle; basically, in the near field the sound pressure level does not have a simple relationship with distance, because of the proximity of vibrating surfaces. In this region levels may remain constant or even increase with distance because of the strong dependence on nearby radiating surfaces.

With respect to noise control, increased absorption on surfaces has some effect on the reverberant field, but none on the direct sound field, which has encountered no reflecting surfaces. For the latter, either direct source modification or the erection of some kind of partition represents a highly effective way of reducing the noise level. Chapter 6 discusses in more detail the approaches to solving these problems.

3.4.4 Sound Power Level Estimation

Sound power is important because it represents a unique characteristic of a particular source which, when properly specified, is independent of the particular environment concerned. This measurement is useful in comparing the relative acoustic properties of various machines and the effects of changing the acoustic conditions on the sound pressure. Knowing the sound power level of a source in appropriate frequency bands, and by making intelligent assumptions about the nature of the sound field generated at points of interest (see section 3.4.3), it is possible to predict accurately the sound pressure level at these points.

The general relationship between the sound power level of a spherically radiating source and the sound pressure level at distance r is derived from the following equation (see section 2.2.3):

$$W = I (4\pi r^2) \quad W$$

Taking logarithms and using the sound pressure level and intensity level equivalence principle (see section 2.4),

$$\begin{aligned} PWL &= SPL + 10 \log 4\pi r^2 \\ &= SPL + 20 \log r + 10 \log 4\pi \\ &= SPL + 20 \log r + 11 \quad dB \end{aligned}$$

where r is the distance from source, in meters.

For a directional source the general equation is given by

$$PWL = SPL + 20 \log r + 11 - DI \quad dB$$

where DI is defined as in section 3.4.2.

In ducts, assuming no losses at the peripheral surfaces, the relationship becomes

$$PWL = SPL_D + 10 \log S_D \quad dB$$

where
SPL_D is the sound pressure level measured just off center line, dB
S_D is the cross section of duct, m^2.

Ideally, the measurement of sound power level should be made either in free-field conditions as in an anechoic chamber, or in a highly reflective reverberant room where a diffuse field may be assumed. For the former case, all the above relationships apply, and it is usual to use a regular array of microphones equidistant from the source and at least $\lambda/4$ from any wall of the room, where λ is the wavelength of the lowest frequency of interest. Figure 3.12 shows a six-point array over a hypothetical hemisphere surrounding a source. The sound power level, PWL, of the source, assuming omnidirectional radiation through equal areas associated with each measuring point, is given by

$$PWL = 10 \log \frac{1}{6} \sum_{i}^{n=6} \left(\frac{p_i}{p_{\text{ref}}} \right)^2 + 20 \log r + 8 \quad dB$$

or

$$PWL = \overline{SPL} + 20 \log r + 8 \quad dB$$

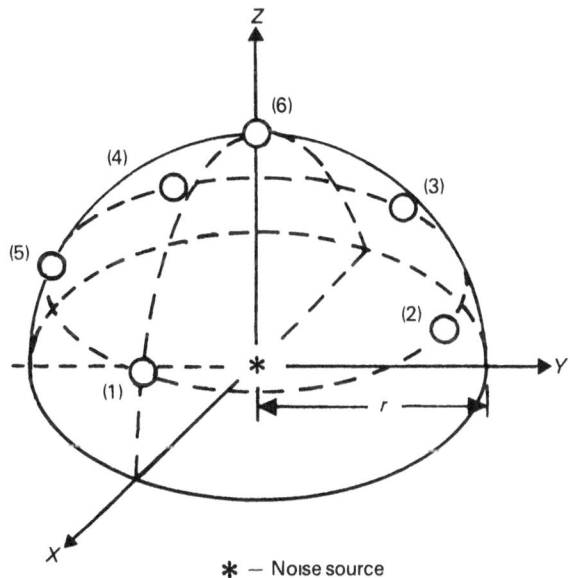

* — Noise source

Coordinates of unit hemisphere			
Microphone position	X	Y	Z
1	0 89	0	0 45
2	0 28	0 85	0 45
3	−0 72	0 53	0 45
4	−0 72	−0 53	0 45
5	0 28	−0.85	0 45
6	0	0	1

Figure 3.12 – Six-point hemispherical microphone array for measuring sound power.

25

where SPL is the logarithmic average of sound pressure levels (space-averaged). Note that the directivity index of hemispherical radiation above the reflecting plane is $+3$ dB (from figure 3.9), so that $PWL = \overline{SPL} + 20 \log r + 11 - 3$.

Typically an even number of microphones is used for this, for example, 4, 6, 10, or 12, depending on the source size and space available. With this method, directivity effects are easily estimated, although if the source is highly directive then a larger number, up to 24 or so, should be used.[26,27] If a sufficient number of measurements of the sound pressure level are taken on the surface of a hypothetical sphere or other regular geometric shape surrounding the source, then the error implicit in failure to sample the level at every point can be minimized.[28] Normally, measurements are made in octave bands and the sound power level for each of these may be summed to give an overall power level.

In a reverberation room, the number of microphones needed in an array, N_M, is generally less than for a free field and is given approximately by

$$N_M \cong \frac{12V}{\lambda^3}$$

where

V is the room volume, m^3
λ is the longest wavelength of interest, m.

Note: The microphones should be placed at least $\lambda/2$ apart and at least twice the maximum source dimension from the source. Their positions are not as critical as for free field assessment, since a diffuse field is being measured.

There is an inherent inaccuracy in assessing pure tone radiation by this method[29]; this error is exacerbated when considering low frequency radiation from large sources.[30] The equation for making an absolute determination under reverberant conditions is[31,32]

$$PWL = \overline{SPL} + 10 \log V - 10 \log T_R + 10 \log \left(1 + \frac{S\lambda}{8V}\right) - 13.5 \quad dB$$

where

$\overline{SPL}$ is the space-averaged sound pressure level, dB
V is the room volume minus source volume, m^3
T_R is the Sabine reverberation time defined in section 6.2.2, s
S is the area of all room boundaries, m^2
λ is the wavelength at frequency of interest, m.

A comparative technique may be employed to determine the sound power level of a test source where the acoustic and geometric characteristics are similar to those of a standard reference source.[33] In this case, the procedure is first to find average sound pressure levels in the appropriate frequency bands for both the reference and test sources. The next stage is to calculate the sound power level of the test source according to

$$PWL_t = \overline{SPL}_t + PWL_{ref} - \overline{SPL}_{ref}$$

where

PWL_{ref}, PWL_t are the sound power levels of reference and test sources, respectively
SPL_{ref}, SPL_t are the space-averaged sound pressure levels of reference and test sources.

In many situations, it is not possible either to obtain prior information about machinery sound power levels or to remove such machinery for laboratory assessment of sound power. In these cases it is necessary to make some kind of in situ measurement, and this is most frequently performed under semireverberant conditions such as exist inside most work places. A thorough review of the different techniques which may be used is given in chapter 4 of reference 26 (see also section 6.3.1). As an example, we outline here the two-surface method developed by Diehl,[26] which is quite suitable for tests on large machinery.

If a 6 dB decrease in sound pressure level is obtainable in a semireverberant field, then even though the distance needed to achieve this is usually greater than it would be in a free field, approximate free field conditions may be assumed. Corrections can be applied to the estimated sound power level in each frequency band to give an approximate free field result. Measurements are made at locations corresponding to specific locations on the surface of a hypothetical parallelepiped enclosing the machine, with total surface area S_1. A second series of measurements is then made on a larger parallelepiped surface S_2, where the major dimensions have the same proportional relationship as for S_1. The average sound pressure levels for each surface, $\overline{SPL}_1$ and $\overline{SPL}_2$, respectively, are then calculated.

It can be shown that the actual sound power is given by

$$PWL = \overline{SPL}_1 + 10 \log S_1 - C \quad dB$$

where the correction factor C is

$$C = 10 \log \left[\frac{E}{1 - E} \left(\frac{S_1}{S_2} - 1 \right) \right]$$

where $E = 10^{(\overline{SPL}_1 - \overline{SPL}_2)/10}$ and S_1, S_2 are surface areas, in square meters.

C is readily found from the family of curves shown in figure 3.13.

Sound power levels must always be quoted for the particular conditions under which the machine is being run, since conditions may change considerably during the course of operation. Also, in some situations, it may not be easy to make far field measurements and the microphones must be placed close to the source. In this case, the measurement surface should conform closely to the source shape, and microphones should be equally sensitive to all incident sound over the front 180° arc.[28]

3.5 SURVEY APPROACH

3.5.1 Initial Preparation

The proper definition of a noise problem is of primary importance; this definition is best achieved by first determining the need for noise control at all. There may be a variety of factors which, either singly or in combination, are responsible for the need for a noise analysis. The most common ones include[33] potential hearing damage to employees because of adverse noise working conditions; determination of whether equipment or vehicles are in compliance with test codes or specifications; community annoyance resulting from external

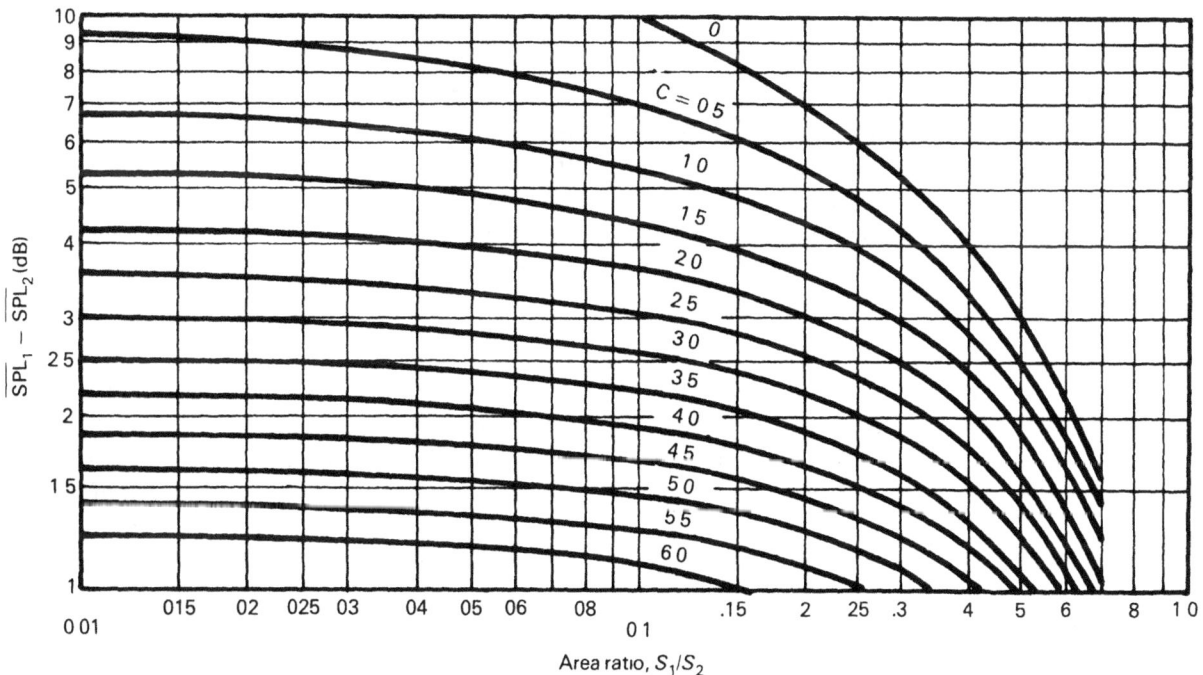

Figure 3.13 – Correction factors for two-surface method of sound power level determination (reference 26). (From George M. Diehl, Machinery Acoustics, 1973. Used with the permission of John Wiley & Sons, Inc.)

noise levels; and noise level contour surveys for general planning purposes. The appropriate criteria for control may then be selected, for example, OSHA noise exposure limits (see the Appendix), community noise ordinances, or sound power level measurement standards. Where possible, a ground plan and evaluation of the building or area concerned should be obtained, as this becomes invaluable in planning surveys and identifying particularly sensitive areas.

Information is also required on the type of noise sources involved, their dimensions, operating characteristics, directionality, and long-term influences such as pattern of use throughout a work week. Particularly important is the need to gain some idea of the level and spectral content of the noise. In this respect, it is emphasized that an empirical on-site investigation even with nothing more than aural inspection by a well-trained acoustician, and careful examination of the above factors, is of considerable value in helping to select the type of equipment required and the measurements necessary. For example, rotating machinery components such as lathes and saws give rise to periodic waveforms, which tend to possess strong discrete frequency components requiring relatively detailed frequency analysis.

3.5.2 Measurement of Acoustic Quantities

Accuracy of measurements is essential. It is imperative to ensure that equipment is functioning correctly and to calibrate it using one of the devices described in section 4.9. Internal reference calibration signals should always be cross checked with external calibrators. Other considerations include the correct selection and deployment of a microphone for the sound field involved, i.e., random incidence, perpendicular incidence, or grazing incidence; and whether avoidance of interference effects when performing precision measurements at high frequencies requires the separate mounting of the microphone or instrument at least 1 m from the observer.

The possibility of instrument overload is reduced by ensuring the utilization of the lowest available measuring range for which the complete range of fluctuations in level still registers on the instrument scale. For sound level meters the ''slow'' response setting (see section 4.4.1) is used frequently unless the sound has a sharply changing nonperiodic amplitude, in which case the ''fast'' response setting is used. Cyclical noises which have

changes in excess of 5 dB should be sampled every 15 seconds or so and the arithmetic average taken after a few minutes. In this case it is also necessary to specify the maximum range of excursions from the mean. Special meter characteristics are required to give an accurate representation of impulsive-type sounds.

Initial measurements should be relatively few in number, provided the level does not vary greatly over large areas. It is usual to obtain representative measurements of a sound field by avoiding any reflecting surfaces by at least 1 m. When assessing personal exposure conditions for individuals, the microphone should be located at the normal position of the head (or 1 m away for in situ measurement). This is taken to be 1.5 m in height for standing positions and 1.1 m for seated workers.[34] When making measurements in a reverberant field, it is normal to take a space-average level, especially for low frequencies.

In general, the measurements are made in dB(A) before further specific analysis into octave, one-third octave bands, etc., is performed. It is always valuable to record a background or ambient level of the sound field when the source of interest is not generating. This permits the effective source contribution to the total noise environment to be estimated and is especially important in dealing with community noise problems. Vibration measurements may also be necessary, particularly when source modifications are being examined or when the relative importance of several structure-borne transmission paths are being considered. This is discussed further in chapter 6. All operating modes of the machines concerned should be accounted for in the survey and analysis.

3.5.3 Data Presentation

Generally, four major areas should be covered: sound field and source description; acoustic and physical description of the environment in which propagation occurs; classification of instrumentation; and operator/receiver positions and exposure conditions.

3.5.3.1 Sound field and source description

Sound field and source description may include the specification of sound power levels as well as full identification of the location using diagrams, sketches, or photographs. Operating conditions should also be noted, including length of running time, cyclical or steady mode, workload required, and so on. Recorded levels should be fully categorized according to measuring position, microphone orientation, weightings used, frequency band, fluctuations, and time.

3.5.3.2 Environment description

A full description of the physical dimensions of the surrounding walls, surfaces, and neighboring equipment is necessary, together with an annotation of their probable acoustic qualities, e.g., absorptive or reflective, massive or lightweight.

3.5.3.3 Instrumentation description

Instrumentation description should include a complete logging of all equipment in terms of manufacturer's type identifications and serial numbers. Frequency bandwidths and signal-to-noise ratios may be specified where relevant. Calibration results before and after the survey should also be included, as well as the instrument settings used for the measurements.

3.5.3.4 Receiver positions and conditions

Description of receiver positions and conditions should include a complete description of levels received at the positions of critical exposure, as well as some account of both temporal and frequency characteristics. Background levels may be included for comparison. Also, a record of machine operator movements throughout a working period should be provided. Where a workman is performing a variety of tasks throughout the day during which his noise environment changes substantially, then it is more relevant to use dosimeters in assessing his personal exposure and to monitor closely his daily routine.

REFERENCES

1. International Organization for Standardization, Normal Equal Loudness Contours for Pure Tones and Normal Threshold of Hearing under Free-Field Listening Conditions, ISO Recommendation R226, 1961.
2. International Organization for Standardization, Method for Calculating Loudness Level, ISO Recommendation R532, 1966.
3. Howes, W. L., Overall Loudness of Steady Sounds According to Theory and Experiment, NASA Reference Publication 1001, 1979.
4. Pearsons, K. S., and R. Bennett, Handbook of Noise Ratings, NASA CR-2376, 1974.
5. Howes, W. L., Overall Loudness Calculation Procedure for Steady Sounds, Cosmic Program Abstract, LEW-12914, 1979.
6. Kryter, K. K., Review of Research and Methods for Measuring the Loudness and Noisiness of Complex Sounds, NASA CR-422, 1966.
7. Beranek, Leo L., *Noise and Vibration Control,* McGraw-Hill Book Co., 1971.
8. Fidell, S., Community Response to Noise in *Handbook of Noise and Control,* Cyril M. Harris, ed., Second ed., Chapter 36, McGraw-Hill Book Co., 1979.
9. Beranek, Leo L., Noise Measurements, Noise Con. Eng., vol. 9, no. 3, 1977, p. 100.
10. Hayes, C. D., and M. D. Lamers, Octave and One-Third Octave Acoustic Noise Spectrum Analysis (JPL TR 32-1052), NASA CR-81251, 1967.
11. ANSI, American National Standard Specification for Octave, Half-Octave, and Third-Octave Band Filter Sets, S1.11-1966 (R1976).
12. Shipley, J. W., and R. A. Slusser, A Digital Technique for Determining Third-Octave Sound Pressure Levels with a More Uniform Confidence Level. JPL TM 33-422, 1969.
13. Ebbing, C. E., and T. H. Hodgson, Diagnostic Tests for Locating Noise Sources: Classical Techniques (Part I), Signal Processing Techniques (Part II), Noise Con. Eng., vol. 3, no. 1, July–August 1974, p. 30.
14. Soderman, P. T., and S. C. Noble, Directional Microphone Array for Acoustic Studies of Wind Tunnel Models, J. Air., vol. 12, no. 3, 1975, p. 168.
15. Kendall, J. M., Airframe Noise Measurements by Acoustic Imaging, AIAA Paper 77-55, 1977.
16. Kendall, J. M., Resolution Enhanced Sound Detecting Apparatus, U.S. Patent 4 149 034, April 1979.
17. Kendall, J. M., Acoustic Imaging System, NASA Tech Brief, Spring 1977, p. 56.
18. Ahtye, W. F., W. R. Miller, and W. C. Meecham, Wing and Flap Noise Measured by Near- and Far-Field Cross-Correlation Techniques, AIAA Paper 79-0667, 1979.
19. Ahtye, W. F., and G. K. Kojima, Correlation Microphone for Measuring Airframe Noise in Large-Scale Wind Tunnels, AIAA Paper 76-553, 1976.

20. Moore, M. T., and E. B. Smith, Sound Separation Probe, NASA Tech Brief B75-10286, 1975.

21. Wilby, J. F., ed., Coherence and Phase Techniques Applied to Noise Source Diagnosis in NASA Ames 7 × 10 Foot Wind Tunnel No. 1, NASA CR-152039, 1977.

22. Brown, D. L., and W. G. Halvorsen, Application of the Coherence Function to Acoustic Noise Measurements, Internoise 72 Proceedings, 1972, p. 417.

23. Piersol, A. G., Use of Coherence and Phase Data Between Two Receivers in Evaluation of Noise Environments, J. Sound Vib., vol. 56, no. 2, 1978, p. 215.

24. Montegani, F. J., Some Propulsion System Noise Data Handling Conventions and Computer Programs used at the Lewis Research Center, NASA TM X-3013, 1974.

25. Montegani, F. J., Some Propulsion System Noise Data Handling Conventions and Computer Programs used at the Lewis Research Center, Cosmic Program Abstract, LEW-12285, 1975.

26. Diehl, G. M., *Machinery Acoustics,* John Wiley & Sons, 1973.

27. ANSI, American National Standard Method for the Physical Measurement of Sound, S1.2-1962 (1971) (partially revised by S1.13-1971 and by S1.21-1972).

28. Holmer, C. I., Procedures for Estimating Sound Power from Measurements of Sound Pressure (prepared by National Bureau of Standards for EPA), EPA-550/8-76-001, 1975. (Available from NTIS as COM-75-11399.)

29. Doak, P.E., Fluctuations of the Sound Pressure Level in Rooms when the Receiver Position is Varied, Acoustica, vol. 9, 1959, p. 1.

30. Smith, W. F., and J. R. Bailey, Investigation of the Statistics of Sound-Power Injection from Low-Frequency Finite-Size Sources in a Reverberant Room, J. Acoust. Soc. Amer., vol. 54, no. 4, 1973, p. 950.

31. Beranek, Leo L., The Measurement of Power Levels and Directivity Patterns of Noise Sources in *Noise and Vibration Control,* Leo L. Beranek, ed., Chapter 6, McGraw-Hill Book Co., 1971.

32. ANSI, American National Standard Methods for the Determination of Sound Power Levels of Small Sources in Reverberation Rooms, S1.21-1972.

33. International Organization for Standardization, Determination of Sound Power Levels of Noise Sources-Methods Using a Reference Sound Source, International Standard 3747, 1975.

34. Wells, Richard J., Noise Measurements: Methods in *Handbook of Noise Control,* Cyril M. Harris, ed., Second ed., Chapter 6, McGraw-Hill Book Co., 1979.

CHAPTER 4

Instrumentation for Sound and Vibration Measurement

To transform the variations in sound pressure associated with audio frequencies (20 Hz to 20 kHz), or the mechanical excitation caused by vibrations in a structure, into an electrical analog signal suitable for processing and subsequent measurement of the original stimulus, a device known as a transducer is required. Those used for measuring sound pressure are known as microphones, of which there are several commonly available types. When making vibration measurements, a transducer known as an accelerometer is most often used.

In both instances, the general processing and analysis system is similar, involving preamplifica-

tion of the relatively weak transducer signal; processing according to predetermined parameters of interest; and eventual display on some form of recorder, meter, or oscilloscope for measurement and assessment by the observer. The schematic diagram in figure 4.1 illustrates typical stages in the measuring chain. Frequently, vibration and sound pressure analyses may be made using the same equipment.

4.1 MICROPHONES

There are currently three types of microphones used for acoustic measurements; table 4.1

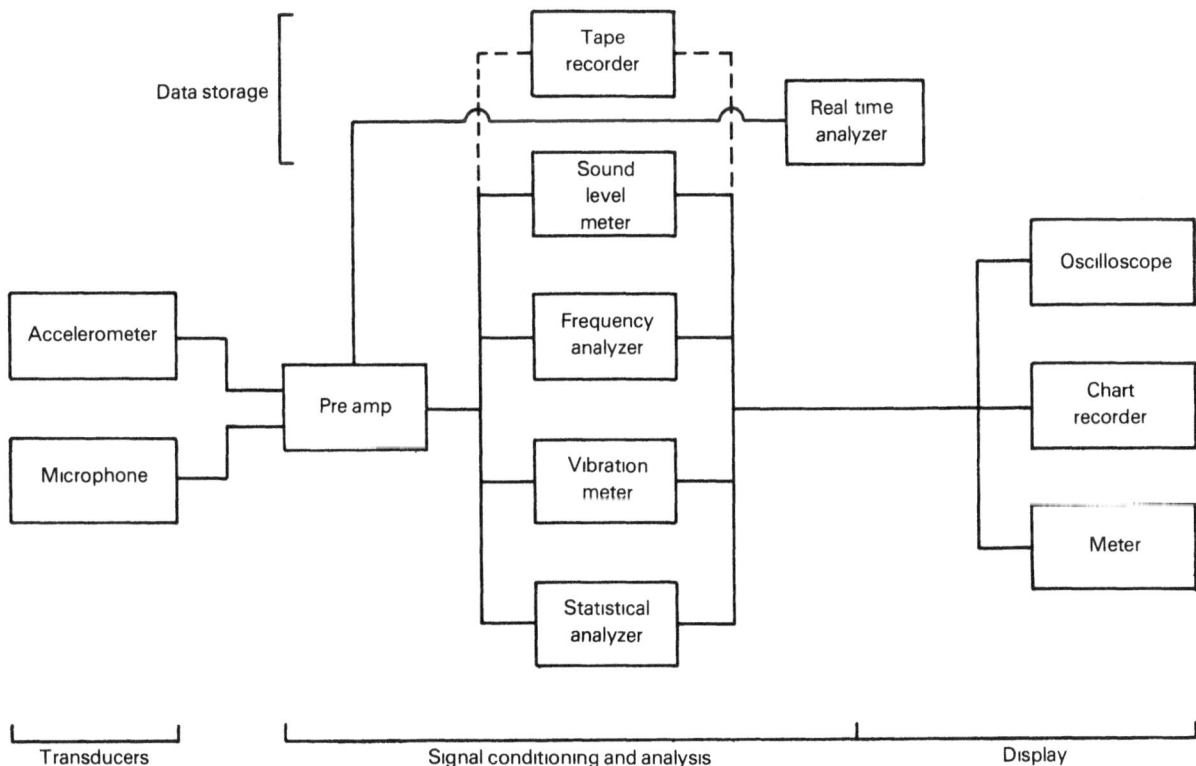

Figure 4.1 – Schematic diagram of typical measuring combinations of sound and vibration equipment.

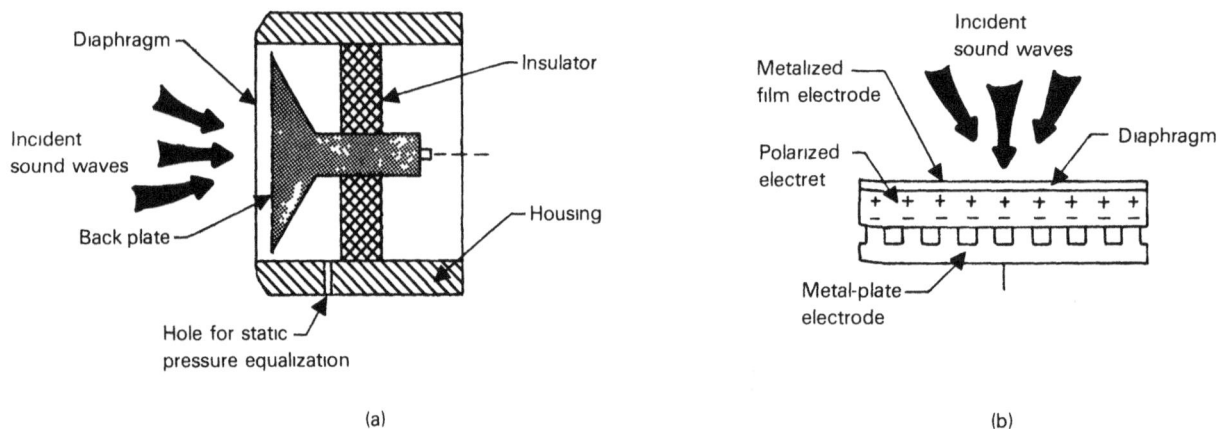

Figure 4.2 - Schematic diagrams of condenser microphones. (a) Air-condenser. (b) Electret-condenser.

Table 4.1 - Characteristic features of microphone types used for acoustic measurements.

	Piezo-electric	Air-condenser	Electret-condenser
Sensitivity	Reasonable	Good	Good
Frequency response	Reasonable	Good	Good
External polarization voltage	None	Yes	None
Long-term stability	Reasonable	Excellent	Good
Influence of humidity	Reasonable	Poor	Good
Ruggedness	Good	Poor	Good
Relative cost	Low	High	Medium

highlights some of the principal features of each. Figure 4.2 illustrates the basic construction details for the more sophisticated condenser types.

4.1.1 Piezoelectric Microphone

This design incorporates a piezoelectric ceramic crystal connected to a diaphragm. This nonconducting crystal produces an electric charge when disturbed by the movement of the diaphragm caused by incident sound pressure.

Piezoelectric microphones do not have as linear a frequency response as the other types discussed below. However, they do have the advantages of lower cost, robustness, and no requirement for a polarization voltage.

4.1.2 Air-Condenser Microphone

This type of microphone has a very thin stretched-metal diaphragm positioned close to a

rigid backplate, with an air gap between them (see figure 4.2a). A polarizing voltage is applied across these, thus forming a capacitor which is highly sensitive to changes in the air gap caused by relative movement of the diaphragm resulting from fluctuating air pressure. The change in capacitance of the device generates an alternating voltage proportional to the sound pressure fluctuations over a large frequency range. There is a small vent which connects the air gap to the ambient atmosphere, so that the diaphragm remains unaffected by changes in the ambient pressure.

Although renowned for their excellent stability and low internal noise, conditions of high humidity adversely affect the output of air-condenser microphones.

4.1.3 Electret-Condenser Microphone

Electret-condenser microphones have become increasingly popular in recent years and have an advantage over the air-condenser type by having a built-in polarizing voltage. This voltage is achieved by incorporating a plastic diaphragm with a metalized sensing surface in front of a prepolarized polymer material (see figure 4.2b). The backplate has a large number of small protuberances, each representing an individual sensing cell. These cells act in concert to provide a composite output proportional to the incident sound pressure. The design is inherently less sensitive to humidity, although at the present time they are not generally believed to possess quite the same long-term stability as their cousins, the air condensers.

4.1.4 Directionality and Sensitivity

The sensitivity of a microphone is measured by the size of its output voltage for a given strength of acoustic signal at the diaphragm. Microphones are extremely sensitive pressure sensors, since for typical industrial situations they need to be able to detect sound pressure in the range 6.3×10^{-4} N/m^2 to 6.3×10^2 N/m^2 (30–150 dB). In order to measure low sound pressure levels they need to have a reasonably large diaphragm. Unfortunately, this conflicts with another prerequisite, which is that the body of the microphone itself should not interfere with the sound field through diffraction or reflection effects, such that the reading obtained is different from one obtained if the microphone were infinitely small. This interference phenomenon predominates at high frequencies, where the wavelength is of the same order or less than the microphone diameter. The net result is to make the output dependent on the angle of incidence of the sound waves. It follows that the smaller the microphone, the more omnidirectional its behavior will be; however, this is accompanied by a reduction in sensitivity. As a guide, 1-inch microphones are directional above about 3 kHz, 1/2-inch above 6 kHz, and 1/4-inch above 12 kHz.[1]

The directionality characteristic, otherwise termed the free field characteristic, is demonstrated by plotting frequency versus relative response in decibels for varying angles of incidence in a free field (see section 3.4.3). This characteristic may be optimized such that linearity through high frequencies is obtained for specified angles of incidence.

Three classes of microphone are normally available: free field or perpendicular-incidence microphones, which have the best response when the diaphragm is perpendicularly oriented with respect to the sound source; random-incidence or pressure microphones, which have the flattest frequency response when the sound field is diffuse and sound waves randomly impinge on the diaphragm at all angles (this roughly corresponds to an average angle of incidence of 70° with respect to the microphone longitudinal axis); and grazing-incidence microphones, which are optimized for best response at 90° angles of incidence. Figure 4.3 illustrates the frequency response characteristics of two classes of microphones and figure 4.4 shows the effect of size on frequency response. Random incidence correctors are available for attaching to free field microphones to change their free field characteristic such that they give good linear response in diffuse sound fields. The American National Standard[2] sets design criteria for random-incidence microphones only.

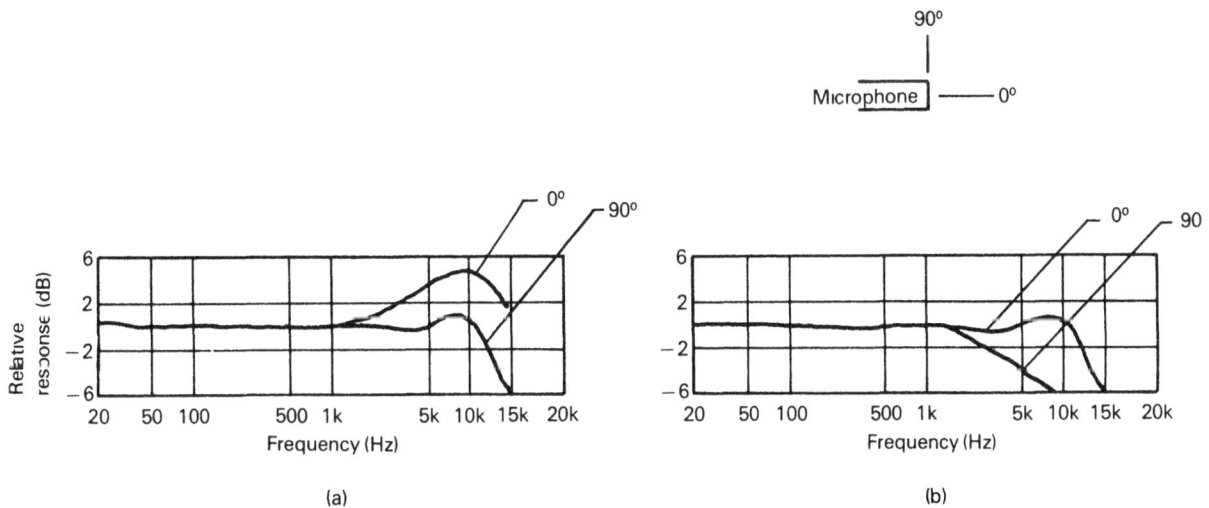

Figure 4.3 – Frequency response characteristics of typical 1-inch microphones. (a) Grazing-incidence microphone. (b) Perpendicular-incidence microphone.

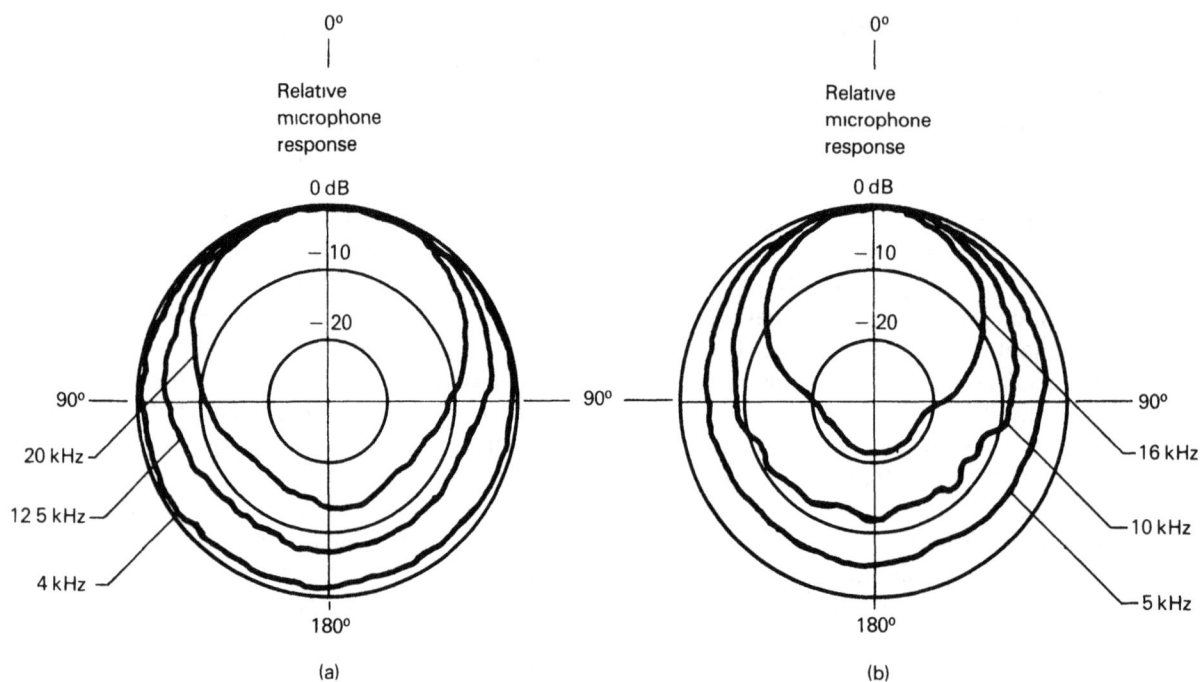

Figure 4.4 – Directivity plots at different frequencies for two sizes of microphones. (a) 1.27-cm (0.5-inch) microphone. (b) 2.54-cm (1-inch) microphone.

4.2 ACCELEROMETERS AND OTHER VIBRATION TRANSDUCERS

Ideally, the location of a vibration transducer should not influence its output; it should perform its function equally well whether it is connected to a structure on the ground, a moving vehicle, or a complex machine. The quantities used to describe vibration movements are either displacement, velocity, or acceleration. The output of the transducer should be directly proportional to the physical quantity of interest and normally it is attached directly to the vibrating body. If this is not feasible, it may be attached to a fixed structure, and measurements made relative to the movement of the body of interest.

Table 4.2 lists the main vibration transducers commonly used. These can be broadly classified into generating or nongenerating types, according to whether an electrical supply voltage is needed; and seismic or nonseismic, depending on whether the device is fixed to the vibrating body or measures relative motion only.

4.2.1 Piezoelectric Accelerometer

Of the devices categorized in table 4.2, the piezoelectric accelerometer is the most popular be-

Table 4.2 – General categorization of vibration transducers.

	Generating	Nongenerating
Seismic	Piezoelectric accelerometer	Strain gauge accelerometer (thin wire or piezoresistive semiconductor construction)
Nonseismic	Electro-dynamic velocity transducer	Noncontacting displacement transducers (utilizing changes in capacitance, inductance, or eddy currents)

cause of its high output relative to size and its relatively uniform frequency response. Generally, it consists of a relatively large mass clamped firmly against two piezoelectric discs which act as a spring between the mass and a solid base. When subjected to vibration, the mass exerts on the discs a force proportional to its acceleration. In turn, the discs generate a voltage proportional to this force and therefore to the acceleration to which the whole unit is exposed.

The sensitivity of the unit depends on its size; however, the larger the unit, the lower its resonant frequency. This latter parameter is important in determining the high frequency performance of

34

accelerometers. In addition, because of its mass, a larger unit will tend to interact more with the response of a system.

Research conducted by NASA and USAF on propellers and fan blades has placed severe constraints on the size and weight of accelerometers. As a major consequence of this work, a series of miniature piezoelectric accelerometers with good response characteristics has been developed by Bolt, Beranek, and Newman, Inc. These possess relatively high sensitivity, built-in preamplifiers, and have a typical sensing diameter of only 0.2 cm. Because of their small size they have also been used to make measurements within thin boundary layers with minimum disturbance to the flow structure. Such miniature accelerometers are of considerable value in any situation where there are severe spatial or mass constraints.

4.2.2 Integrators

Electronic integrators, inserted in series between a transducer and the preamplifier (see section 4.3), make the output of an accelerometer frequency sensitive. By switching in the appropriate frequency slope characteristic, an output proportional to displacement, velocity, or acceleration can be produced. These quantities can therefore be measured directly, extending the versatility of piezoelectric devices.

4.3 PREAMPLIFIERS

Preamplifiers are used to provide impedance matching and to boost the relatively low output signal from the transducer before it is routed through signal conditioning systems.

Voltage and charge amplifiers are two types of preamplifiers in common use. Condenser microphones or piezoelectric transducers may be regarded as having a constant charge sensitivity. This means that a long length of cable between microphone and preamplifier, which increases the capacitance of the system, reduces the output voltage. Hence voltage preamps must always be used close to the transducer. Some preamplifiers provide a polarization voltage for use with air-condenser microphones.

On the other hand, charge amplifiers, which are sensitive only to charge, may be located at any distance from the transducer. Charge amplifiers are, however, more expensive than voltage preamplifiers.

4.4 SOUND LEVEL METERS

Sound level meters represent a complete portable measuring system in one unit. They are especially convenient for field use and can be adapted to take input from an accelerometer so that vibration measurements may be made directly. There is a great variety of instruments available on the market and a wide variety of applications to which they may be put, from making sound power measurement surveys to monitoring of factory ambient noise levels.

4.4.1 Electrical and Operational Characteristics

The basic components of a sound level meter are shown in figure 4.5. Because of the large range of sound levels (up to 30–150 dB) which the instrument may be required to measure, representing as much as 10^6 change in pressure and hence signal

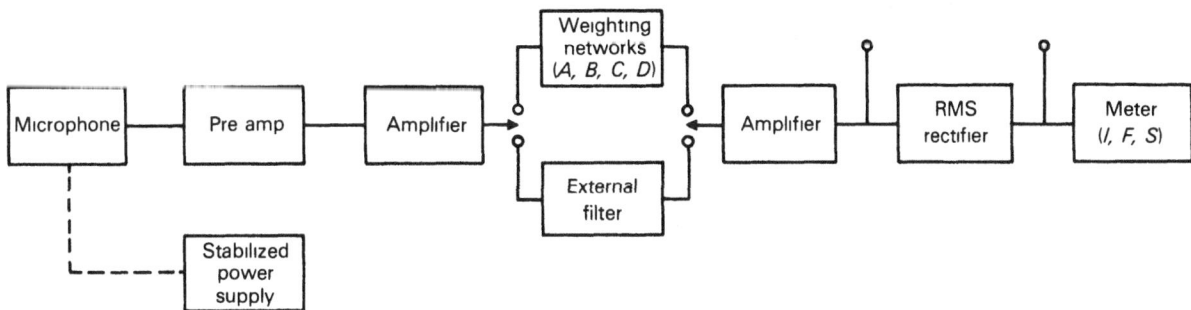

Figure 4.5 – Schematic diagram showing component parts of a sound level meter. Response characteristics of meter needle with time constants in parentheses may be either I, *impulse (35 ms);* F, *fast (125 ms); or* S, *slow (1000 ms).*

magnitude, the amplifier inputs are controlled by fixed attenuators, usually in steps of 10 dB. More recently, with the advent of digital processing techniques, it has become possible to extend the scale over which a meter can measure without changing the attenuation.

Electronic weighting scales, discussed in the previous chapter, are incorporated to assess the effect of noise on people. Usually the A-weighted scale is employed (see section 3.1.2). Many instruments also have facilities for switching in a frequency filter which may be either octave or one-third octave band.

The meter needle response is determined by preselection of a fast, slow, impulse, or peak hold characteristic. These characteristics determine the time over which the RMS signal from the rectifier is integrated to provide an averaged reading. The fast and slow modes incorporate 1/8-s and 1-s time constants, respectively. The peak mode has a very short time constant, less than 50 μs, to assess short duration sounds such as punch press or impact noise. It has a rapid response to increasing sound levels and a slower response to decreasing levels. The proposed new OSHA regulations (see the Appendix) specify the requirements for this. The impulse mode with a 35-ms time constant is more commonly used outside of the United States. Recently, the detection-averaging characteristic of sound level meters has been significantly improved, especially with the recognition of danger to hearing from impulsive sounds, which have a high instantaneous peak level relative to the RMS level. The crest factor, which is the ratio of the pressure associated with these two measurements, has been extended in many impulse precision meters to a factor of 20 dB or even more, so that when taking measurements of impulsive sounds there is less danger of the meter's failing to give a true indication of the sound level. Additionally, some meters have an overload indication to give a warning when the signal reading may be in error.

Differential sound level meters can provide a direct indication of small differences in sound pressure between two sites. This may be of advantage when estimating insertion loss characteristics across partitions or spatial variations in level within a sound field. One instrument whose development was sponsored by NASA[3] is especially suitable for measuring high intensity sounds; it features automatic gain control in order to ensure the continuous matching of gains on both channels as the signals change.

Readings may be taken by eye-averaging where the total range of fluctuation is within ±3 dB or so, and the total range of variation is also recorded. For steady sounds, the "slow" response mode may be used, but for short-term variations, such as the cycle of a cutting machine, the "fast" response is more appropriate. In order to take even longer-term averages, L_{eq} readings may be made (see section 3.1.3), since these can be integrated over any time period. AC and DC outputs are often provided so that tape recordings of the measured sound may be made or amplitude analysis in either the frequency or time domains performed.

4.4.2 Standards and Specifications

The principal discriminating feature in terms of quality and accuracy of different sound level meters is the type of microphone used and its characteristics. There are now three international standards,[4-6] which are soon to be replaced and updated by a single document, and a single national one,[2] which is also scheduled to undergo revision. The amended standards will recategorize instruments in terms of four classifications. Even though the design goals for all the standards are identical in regard to frequency response, the tolerances allowed vary considerably at low and high frequencies. The American standard specifies greatest accuracy for random incidence, while IEC requirements are for optimum performance at either 0° or 90° incidence.

The three major classes of sound level meters in current use are discussed in more detail below with an indication of their use.

The Type 3–Survey Meter is rarely used except as a quick check on noise levels. Accuracy[2] may only be ±3 dB at 1 kHz, when A-weighted, but it does offer the advantages of economy and compactness.

Type 2–General Purpose Meters often use condenser microphones and have an accuracy of at least ±2 dB at 1 kHz. The A, B, and C weighting networks (see section 3.1.2) are usually incorporated and measurements may be made over a wider range of sound levels than with Type 3 meters. Special meters known as Type S2A offer only A-weighted measurement capability and are relatively inexpensive. Occasionally, octave-band analyzers are also incorporated as well as detachable microphones. Examples of Type 2 are illustrated in figure 4.6a.

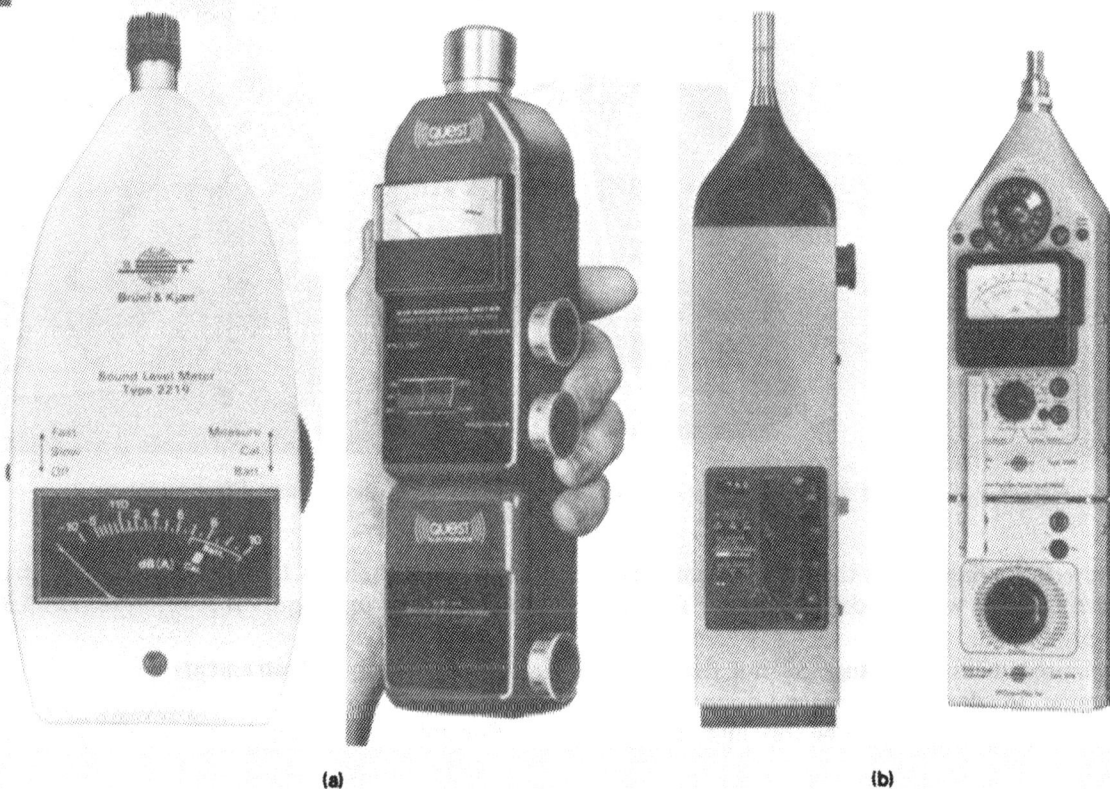

Figure 4.6 - Sound level meters. (a) General-purpose sound level meters (Type 2) (by courtesy of Bruel & Kjaer and Quest Electronics). (b) Impulse and precision sound level meters (Type 1) (by courtesy of GenRad and Bruel & Kjaer).

Type 1-Precision Meters are used only in carefully controlled field environments or in laboratory work. They have accurate microphones of either the air- or electret-condenser type, which may be mounted remotely in order to minimize the effect of the observer and instrument case on the incident sound field. Type 1 meters measure over a larger range than the other two types and have an accuracy of at least ±1 dB at 1 kHz when A-weighted.

Impulse precision meters are governed by IEC 179A.[6] These normally have all the facilities of a Type 1, with an additional impulse and sometimes peak hold characteristic. They can measure signals with a high crest factor in any mode and possess an overload signal indicator, which may be important when dealing with complex signals.

4.5 NOISE DOSIMETERS

Noise dosimeters are commonly used to determine an individual's exposure to noise in industrial situations. They integrate noise exposure over time periods of several hours or more, so as to make a comparison with OSHA specified limits (see the Appendix). Generally, they read out results as a fraction or percentage of total allowable dosage; sometimes a separate readout unit is required, to prevent manipulation of the received dose by the wearer. Typically, dosimeters have a restricted range over which they register, often 90-115 dB(A), although a detector may store any signals in excess of this range. OSHA limits are exceeded if the noise exposure index obtained from summation of levels over an 8-hour day is greater than 1.0.

Noise dosimeters provide a more reasonable estimate of an individual's complete noise exposure than representative measurements taken with a sound level meter. This is particularly true when the operator's position may change during the day or when the operator is exposed to varying noise levels. Dosimeters have been designed as compact

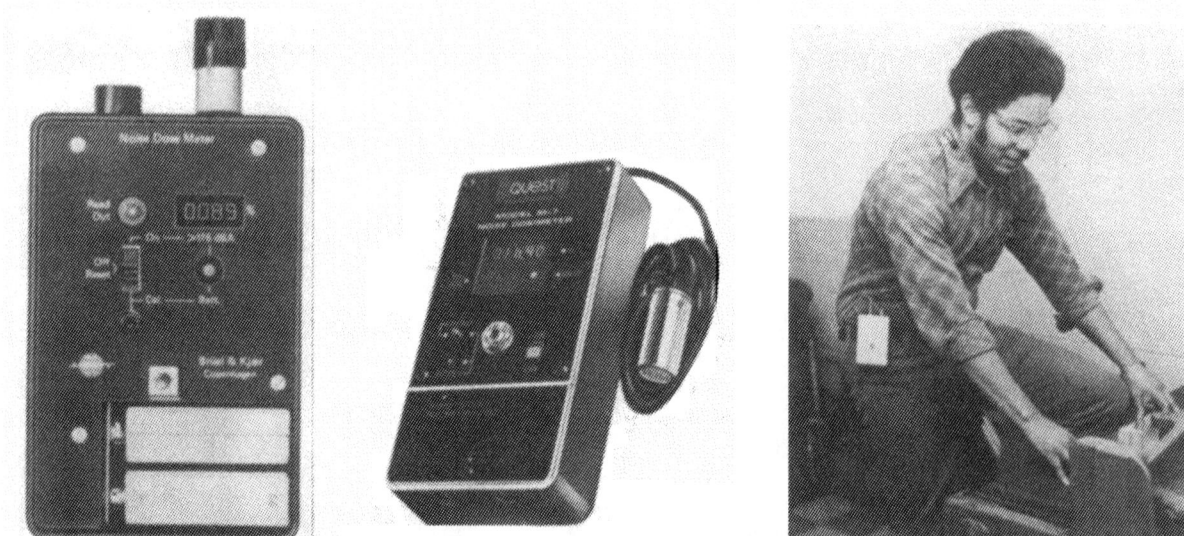

Figure 4.7 - Noise dosimeters (by courtesy of Bruel & Kjaer and Quest Electronics).

and lightweight units so that they can be comfortably worn during a working day. Figure 4.7 shows some typical units.

A limited measuring range[7] is not the only shortcoming in dosimeters. In addition, their operating crest factor may not be very high, limiting sensitivity to impulsive-type sounds. In addition, their accuracy in measuring any particular level may be less than that of a sound level meter. Also, the microphone must be positioned correctly on either the shoulder lapel or breast pocket, according to the type of sound field involved,[8] and the wearer must be instructed appropriately. Accurate calibration is essential, because of the lengthy measurement periods usually involved (see section 4.9.1), and the complete instrument must be protected from adverse environments such as dust and heat.

4.6 FREQUENCY ANALYZERS

Frequency analysis is an important feature of noise control engineering, especially when designing enclosures, optimizing absorption linings, calculating sound power levels of machinery, or identifying noise sources within a complex sound field. All of these procedures, as well as noise control materials, exhibit a strong frequency dependence. A single number assessment of the frequency spectrum, such as comparing *A, C,* and linear-weighted readings to give an approximate assessment of low or high frequency content, is not sufficient for establishing design and control criteria (see section 3.3.1). A full discussion of frequency analysis techniques is given in section 3.3.

4.6.1 Serial Analysis Instruments

Portable serial analysis instruments (see section 3.3.2) in the form of fixed percentage bandwidth frequency analyzers complementary to sound level meters are now commonly available. These are usually octave band or one-third octave band units, they enable analyses to be made conveniently on site. Fixed bandwidth instruments, such as 20 Hz, 10 Hz, and 5 Hz, have until recently been too bulky to use in the field because they employ many electronic filters; and it has often been necessary to tape record the signal for future laboratory analysis. Real time analyzers (see section 4.6.2) have largely supplanted such fixed bandwidth analog instruments.

A complete analysis with a serial frequency analyzer is a lengthy process, and the results are invalid if the signal spectrum or level changes during the analysis. Figure 4.8 shows a typical frequency analyzer of this type.

4.6.2 Real Time Analyzers

Within the last decade or so, the development of high-speed analog-to-digital converters, integrated circuitry, and increased digital memory capacity has enabled frequency analyses to be completed in a matter of seconds. This technique is called real time analysis, and it is clearly an

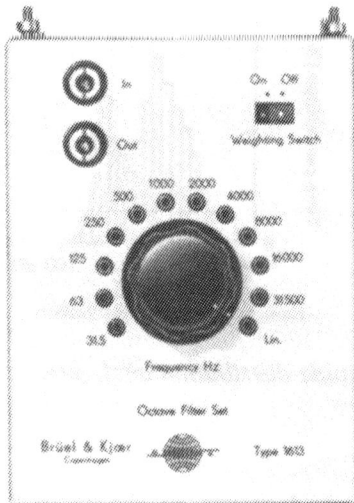

Figure 4.8 - Serial frequency analyzer (by courtesy of Bruel & Kjaer).

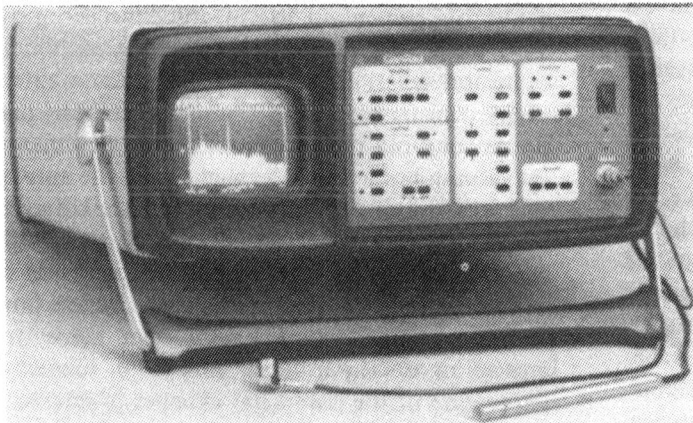

(a)

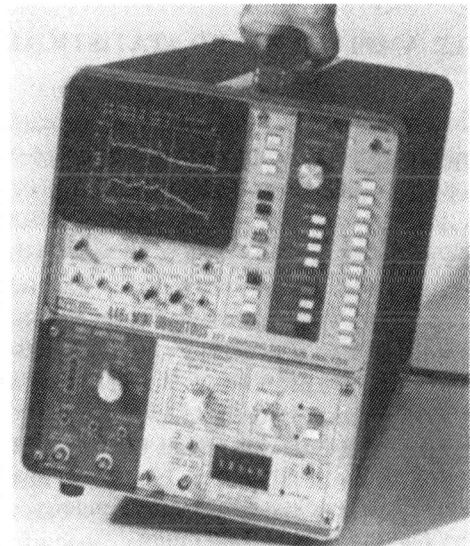

(b)

Figure 4.9 - Portable real time spectrum analyzers. (a) By courtesy of GenRad. (b) By courtesy of Nicolet Scientific Corporation.

advantage over the serial process of manually switching from one filter to another, since essentially all frequency bands are examined at the same time. The rate at which digital sampling of the analog signal occurs predetermines the real time frequency range, although it may be considerably below the highest useful frequency range of the instrument. For example, an instrument capable of analyzing from 0 to 20 kHz may only have true real time capability from 0 to 2 kHz.

Real time analyzers can perform various functions,[9] and the popular fast Fourier transform (FFT) type computes power spectral densities, correlation functions, and the like (see sections 3.3.1 and 3.4.1). The resolution of the instrument is determined by the number of lines on the display screen, representing the number of constant bandwidths available. Fixed percentage bandwidth capabilities may also be provided, and the analyzers are especially useful for examining transient sounds.

Clearly, then, these instruments represent a very useful and versatile tool to the noise control engineer. Although they are becoming lighter, cheaper, and easier to use, the serial type filter will probably remain in popular usage in the field for a considerable time for reasons of economy and simplicity. Figure 4.9 shows some typical real time analyzers.

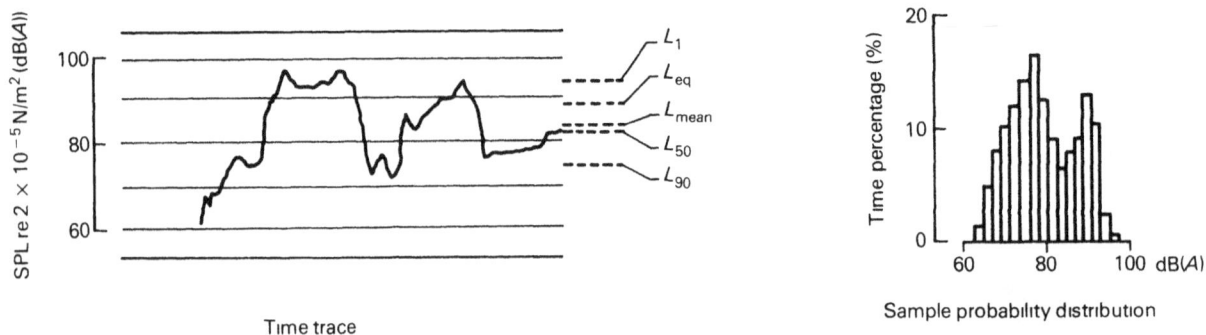

Figure 4.10 – Typical traces and parameters obtainable from amplitude distribution analyzers.

4.7 AMPLITUDE AND STATISTICAL ANALYZERS

Many signals analyzed in noise control have a fluctuating level which may be described in a number of different ways. An energy average taken over a specified period of time constitutes an L_{eq} measure, and the fast or slow responses on sound level meters constitute short-term RMS averaging. The time-varying facet of a noise environment may also be more precisely described by recording the proportion of time during which the sound occurs at each of a series of contiguous sound level intervals within the range of interest. With recent developments in analysis instrumentation, portable digital units have become available. These can quickly give a wide range of statistical parameters, from probability distributions to the sound level that is exceeded $X\%$ of the total sampled time (usually designated L_X).

A typical selection of statistical parameters and their relationships is illustrated in figure 4.10. Although the total number of samples which may be taken is limited, the integrating period of the analyzer may be reduced to permit finer categorization of a highly fluctuating signal, or lengthened to allow a larger total measuring time.

4.8 RECORDERS

Recorders retain a convenient record of the analysis signal which can be conveniently transposed, reexamined, or referenced at a later date. Figure 4.11 shows examples of both graphic and magnetic tape recorders.

4.8.1 Graphic Recorders

Graphic recorders are generally of the strip chart variety: a continuous roll of paper is fed past a recording pen which marks on the paper the signal parameter of interest. This measure is usually either amplitude against time or frequency analysis in percentage bandwidths. Specific applications of these instruments include measurement of sound level with time, reverberation times, sound transmission loss values for insulating partitions, and frequency calibration of instruments. By varying both pen and chart speed, the equivalent response of a sound level meter needle may be modeled. Graphic recorders are usually laboratory based, although portable recorders which may be used in conjunction with sound level meters have become available. These recorders allow the analyst to obtain a quick and clear subjective impression of the parameter changes of interest.

4.8.2 Magnetic Tape Recorders

Magnetic tape recorders store a magnetic analog of sound pressure or mechanical vibration response so that subsequent analysis may be performed at a more convenient time and often with greater sophistication than at the recording site. Moreover, the capability of recording several signals simultaneously can help in making sound power level and survey-type measurements. Also, by changing the tape speed, signals with wide bandwidths may be brought within the scope of the analysis equipment. Digital signal processing increasingly obviates the need for tape recorders in many of these roles.

Frequently, a recorder is the weak link in a measurement chain; direct on-line measurement is generally preferred if possible. When selecting recorders, consideration must be given to input monitoring, distortions imparted to the original signal such as phase-shift, adequate signal-to-

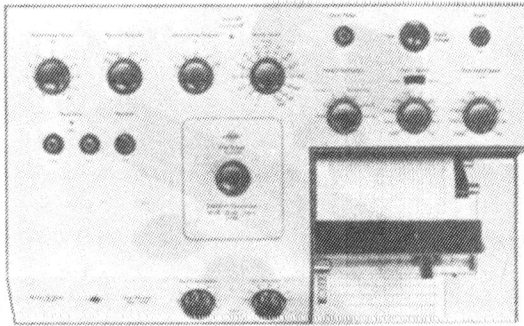
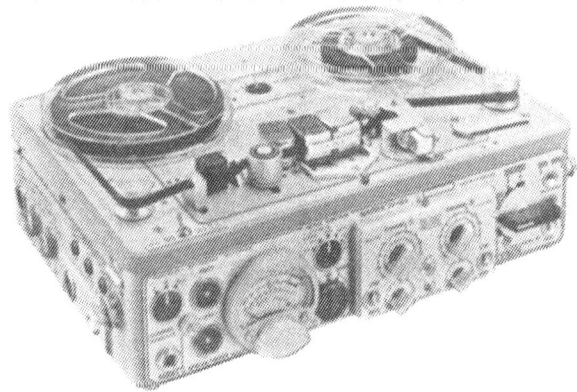

Figure 4.11 - Recorders used in noise and vibration analysis. (a) Graphic level recorder (by courtesy of Bruel & Kjaer). (b) Magnetic tape recorder (by courtesy of Nagra Kudelski).

noise ratio, calibration of the taped signal, and compatibility of the tape with the recorder.

Magnetic tape recorders are either direct or frequency-modulated. Within the former, professional hi-fi standard is normally satisfactory using one-quarter inch tape, but instrumentation class equipment with a larger channel capacity may also be used. These are adequate within the audio frequency range of 40 Hz to 14 kHz. FM recorders are usually used for vibration studies, or for situations where low frequencies from 50 Hz down to DC must be recorded, or where good phase representation is required, as with impulse waveforms.

4.9 CALIBRATION

In order to get maximum accuracy from a measuring system the measurement must be compared against a calibrated signal. This ensures the reproducibility of the measurement and, if properly carried out, gives correct absolute values. A system can be accurately calibrated by applying a reference source, which produces a known level at a specified frequency, to the sound or vibration transducer and adjusting the sensitivity until the RMS value indicated by the meter agrees with the reference. This calibration should be performed immediately before and after measurement on location.

It is also possible to calibrate by feeding an internal reference signal to the preamplifier; however, the accuracy of this method depends on the absolute calibration of the internal reference signal.

4.9.1 Sound Pressure Calibrators

Pistonphones, electrically generated sources, and falling ball calibrators are all used for calibrating sound pressure readings in measurement equipment. The first two are more accurate because the calibrator is fitted securely around the microphone neck with a known void in between. Examples of both are shown in figure 4.12. There are also electronic calibrators that produce known sound pressure levels at several different frequencies, thus allowing the frequency and weighting network responses to be checked. Pistonphones have a cam-driven diaphragm which produces a known sound level at a single low frequency. Falling ball calibrators rely on a collection of ball bearings which pass through an orifice and strike a taut skin diaphragm. The constriction prevents all the balls from falling at once. These units only give an approximate calibration; over a period of time they degrade because the diaphragm stretches.

Dosimeters are calibrated with a stop watch in conjunction with a standard sound level meter calibrator suitably coupled to the dosimeter. The level used must be higher than 90 dB to register on the dosimeter, since these are specifically designed to assess compliance with OSHA limits, which currently apply only to levels in excess of this value.

4.9.2 Vibration Calibrators

Vibration calibrators subject the transducer to a known vibration level and the measuring equipment sensitivity is then suitably adjusted. The value used most commonly is a peak acceleration of 1 g at a single frequency, usually 80 or

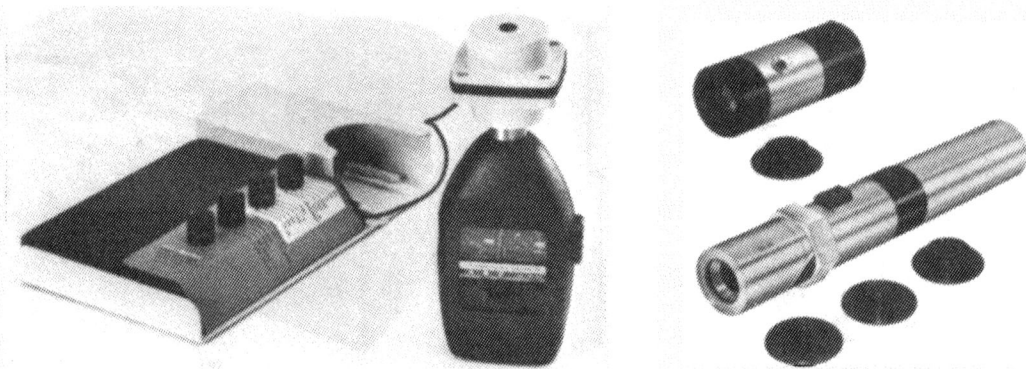

Figure 4.12 – Field sound level calibrators and adaptor rings for various microphone sizes (by courtesy of GenRad and Bruel & Kjaer).

100 Hz (g is unit acceleration from gravity, or 9.81 N/m^2).

4.10 MISCELLANEOUS ANCILLARY EQUIPMENT

The most important items of ancillary equipment are those which improve the performance of microphones, which are very sensitive to adverse environments.[10] For example, dehumidifiers are available as a screw-on connection to air-condenser microphone cartridges for high humidity environments. Push-on windscreens can be used for outdoor noise measurements and are effective for wind speeds of up to around 4 m/s. Nose cones are used for reducing turbulence noise in high speed, relatively uniform flow air streams, such as in ventilation or exhaust ducts. Because the acoustic signal must be separated from within the highly turbulent air flow in wind tunnels and ducts, NASA and other organizations have developed a variety of microphone attachments, the best of which suppress turbulent noise by 12–15 dB relative to nose cone performance.[11,12] These may take the form of porous surface airfoils and porous or slit tubes, with a well-defined directionality characteristic. Tripods and extension cables also reduce the interaction between the acoustic field and the instrument case and observer.

REFERENCES

1. Schneider, Anthony, Instrumentation Microphones, Sound and Vibration, March 1977, p. 6.
2. ANSI, American National Specification for Sound Level Meters, ANSI S1.4-1971.
3. Zuckerwar, A. J., and R. E. Jones, A Differential Sound Pressure Level Meter, Third Interagency Symposium on University Research in Transportation Noise, November 1975, p. 105.
4. International Electro-technical Commission, Recommendations for Sound Level Meters, Publication 123, 1961.
5. International Electro-technical Commission, Precision Sound Level Meters, Publication 179, 1973.
6. International Electro-technical Commission, Additional Characteristics for the Measurement of Impulsive Sounds, Publication 179A (First Supplement to Publication 179), 1973.
7. Leasure, William A., Performance Evaluation of Personal Noise Exposure Meters, Sound and Vibration, March 1974, p. 36.
8. Seiler, John P., Noise Dosimeters, Sound and Vibration, March 1977, p. 18.
9. Kamperman, George W., and Mary Moore, Real Time Frequency Analyzers for Noise Control, Noise Control Engineering, vol. 9, no. 3, Nov.–Dec. 1977, p. 131.
10. Magrab, E. B., Environmental Effects on Microphones and Type II Sound Level Meters, National Bureau of Standards, Technical Note 931, 1976.
11. Noiseux, D. E., Porous Surface Microphone for Measuring Acoustic Signals in Turbulent Windstreams, NASA Tech. Brief B73-10490, 1973.
12. Crocker, M. J., R. Cohen, and J. S. Wang, Recent Developments in the Design of Tubular Microphone Windscreens for In-Duct Fan Sound Power Measurements, Inter-Noise 73 Proceedings, 1973, p. 594.

CHAPTER 5

Noise and Vibration Control Materials

This chapter describes classes of noise and vibration control materials, tabulates their specifications, and outlines their general application. In theory, any material has some effect on the propagation of acoustic energy, but we only review here those commonly used for modifying the acoustic environment to improve the exposure conditions of people at their place of work.

Such materials work by redirecting acoustical and vibrational energy or converting such energy to another form, typically thermal energy. The various modes of operation involved and the relationship of air- or fluid-borne acoustic energy to structure-borne energy are shown in figure 5.1.

The difference between the incident and reflected acoustical energy $(I_1 - I_2)$, divided by the incident intensity, I_1, is a measure of the total degree of absorption, α, which takes place at that boundary. The residual energy intensity $(I_1 - I_2)$, which is not reflected, may either be dissipated within the structure itself by friction, viscous, and hysteretic losses (total intensity I_3), or transmitted away from the partition by structure-borne vibra-

tion (total intensity I_3'). The remainder is re-radiated at the air–structure boundary of the partition with intensity I_4. The degree of re-radiated energy on the receiver side of the structure is characterized by the transmission coefficient, τ. τ is numerically equal to the ratio of transmitted intensity, I_4, to the incident intensity, I_1. Because energy is always conserved, these intensities are related thus:

$$I_1 = I_2 + I_3 + I_3' + I_4$$

The coefficients α and τ, numerically defined in figure 5.1, describe the ability of a material to absorb and transmit, respectively, air- or fluid-borne acoustical energy, and each has a maximum value of unity. The loss factor, η, is used to quantify a material's ability to attenuate structural vibration by internal damping. Certain materials are frequently applied to vibrating surfaces in order to reduce the radiated sound energy. This is distinct from vibration isolation, which is the process by which a material is used to inhibit the transmission of vibrational energy from one structure to another.

The four major categories of acoustic materials are absorbers, insulators, damping materials, and vibration isolators. Absorption materials are normally lightweight, and fibrous or highly porous, whereas effective barrier or insulating materials are of high density and often quite limp. Hence, in many noise control applications, absorption and insulation materials often are used together, since their distinctly different functions and modes of operation make it difficult to attenuate effectively the sound at a receiver by the use of either material on its own. Damping materials usually consist of viscoelastic materials such as epoxy–polymer mixes or rubbers, which are inserted or adhered onto machine parts, panels, and so on, in a number of possible configurations. There is a wide variety of vibration isolators in use; they typically

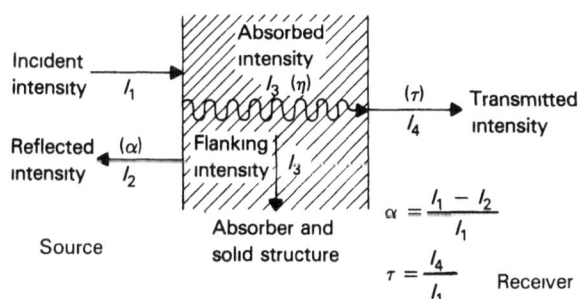

Figure 5.1 – Relationship of acoustic energy intensities for propagation across a solid structure. The acoustic coefficients with most influence on relevant intensities are shown in parentheses. Strictly speaking, this conceptual illustration is only relevant for normal propagation to the partition.

consist of lightly damped steel springs, elastomeric rubber compounds, or even pneumatic devices, which must all possess sufficient resilience to support the loads required.

The following sections outline the means of optimizing the specification for each of these four categories of materials and the basic quantities which affect their performance.

5.1 ABSORPTION MATERIALS

5.1.1 Physical Theory

The mechanism of absorption involves conversion of the kinetic and potential energy of incident acoustic waves to thermal energy. This may occur either via viscous losses due to oscillatory motion of air molecules in absorption materials such as porous foams, or via vibration of the entire material fabric, which may be the solid skeleton of closed-cell foams, or impermeable panels at resonance, as in wall panel absorbers. Fibrous materials mainly absorb energy through the effect of bending and friction between the fibers themselves.

For effective absorption, the pores and voids in cellular materials must be highly interconnected. This is commonly referred to as the degree of reticulation. The specific parameter normally measured in connection with this quality is known as the flow resistance, of which a crude assessment may be made by judging the ease with which one can blow air through the material. If the resistance is too low, little energy conversion can take place within the material; too high a resistance limits air motion and hence reduces oscillatory and friction losses. An interrelated quantity also of interest is the degree of porosity in the material, usually simply related to the density. A third quantity, also of importance in describing the behavior of absorbers, is the structure factor, which accounts for detailed inner conformation and describes the effective change in fluid density which occurs within the absorber voids.

Characteristically, it has been rather difficult to predict the performance of the so-called bulk absorber materials without relying heavily on empirical techniques. The recent interest of the aircraft industry and NASA in the deployment of bulk absorbers in jet engine nacelles has led to a refinement in the analytical prediction of absorber behavior,[1] especially as a result of experiments with highly porous materials with a very uniform composition such as Kevlar and Scottfoam.[2,3] A reasonable prediction of the normal surface impedances (see section 2.2.5) and normal absorption coefficients was achieved by identifying the relevant structure factor and introducing a new parameter called the quality factor. The particular effect of all these parameters on the absorption properties of a material are not discussed further here since the relationships are often complex and do not always match empirical data well. The user is directed to chapter 10 of reference 4 for further information on the subject.

The absorption coefficient, α, is the single most important macroscopic quantity of interest; it is defined variously according to the mode of measurement used. It is nearly always highly dependent on the frequency of the incident sound wave. A standing wave tube is used to measure the normal incidence absorption coefficient, α_n. This measurement is made with a single frequency sinusoidal wave input by a loudspeaker, with a disc of the material under test placed at right angles to the direction of sound propagation at the other end of the tube. Although this technique is suitable for comparing the properties of absorbers quickly, it does not provide values which adequately match the typical performance in practical installations. To give a practical measure, a panel is inserted into a reverberation chamber and the absorption coefficient calculated as a function of the change in the reverberation time (see section 6.2.2). This case deals with a diffuse sound field where the sound waves are incident from approximately all angles at the material surface. This approach enables a statistical absorption coefficient, α_{stat}, to be calculated. The noise field is usually generated by a random broad-spectrum source; quoted specifications must correspond to measurement by this technique as specified in the standard testing procedure.[5] Unless otherwise indicated, all future references to absorption coefficients will implicitly denote use of α_{stat}. Table 5.1 gives some typical absorption coefficients for a selection of noise control materials.

Since the absorption coefficient is a measure of the fraction of acoustical energy which enters the material and is not reflected, a perfect absorber would have $\alpha = 1$. Any practical absorber should have a coefficient of at least 0.60, corresponding to 60% absorption. This represents a reduction in acoustic intensity at the boundary of only 4 dB ($10 \log (0.6/1) = 4$ dB). Normally, absorption coefficients are quoted for octave bands between

Table 5.1 – Sound absorption coefficients α for some typical noise control materials.

Material	Octave band center frequency, Hz					
	125	250	500	1000	2000	4000
Brick, unglazed	0.03	0.03	0.03	0.04	0.04	0.05
Brick, painted	0.01	0.01	0.02	0.02	0.02	0.02
Concrete	0.01	0.01	0.02	0.02	0.02	0.03
Concrete block, painted	0.10	0.05	0.06	0.07	0.09	0.08
Linoleum, rubber, or cork on concrete	0.02	0.03	0.03	0.03	0.03	0.02
Wood	0.15	0.11	0.10	0.07	0.06	0.07
Glass						
<4 mm thick	0.35	0.25	0.20	0.10	0.05	0.05
6-mm plate	0.15	0.06	0.04	0.03	0.02	0.02
Plaster						
On solid surface	0.01	0.02	0.02	0.03	0.04	0.05
On lath	0.10	0.10	0.06	0.05	0.04	0.03
Plywood, 10 mm thick	0.30	0.20	0.17	0.10	0.10	0.11
Gypsum, 12.7 mm thick on 51 × 102-mm studs 410 mm apart	0.29	0.10	0.05	0.04	0.07	0.09
Fiber glass						
25.4 mm thick, 24–48 kg/m^3	0.08	0.25	0.65	0.85	0.80	0.75
51 mm thick, 24–48 kg/m^3	0.17	0.55	0.80	0.90	0.85	0.80
25.4 mm thick, 24–48 kg/m^3, 25.4-mm air space	0.15	0.55	0.80	0.90	0.85	0.80
51 mm thick, plastic covering with perforated metal facing	0.33	0.79	0.99	0.91	0.76	0.64
Polyurethane foam						
6.3 mm thick	0.05	0.07	0.10	0.20	0.45	0.81
12.7 mm thick	0.05	0.12	0.25	0.57	0.89	0.98
25.4 mm thick	0.14	0.30	0.63	0.91	0.98	0.91
51 mm thick	0.35	0.51	0.82	0.98	0.97	0.95
Mineral fiber						
12.7 mm thick	0.05	0.15	0.45	0.70	0.80	0.80
25.4 mm thick	0.10	0.30	0.60	0.90	0.90	0.95
12.6 mm thick on lath, 25.4-mm air space	0.25	0.50	0.80	0.90	0.90	0.85
Drape						
0.35 kg/m^2, hung straight against the wall	0.03	0.04	0.11	0.17	0.24	0.35
0.61 kg/m^2, draped to half area	0.14	0.35	0.55	0.72	0.70	0.65
Carpet						
Haircord on felt	0.05	0.05	0.10	0.20	0.45	0.65
Thick pile on sponge rubber	0.15	0.25	0.50	0.60	0.70	0.70

125 and 4000 Hz. A single-number rating, known as the noise reduction coefficient (NRC), is often taken; the NRC is the arithmetic average of the co-efficient values in the octave bands 250, 500, 1000, and 2000 Hz. This is not useful at extreme ends of the frequency range or for situations where a high degree of optimization at a particular frequency is called for.

Theoretically, maximum absorption occurs when the absorber thickness is about one-fourth the wavelength of the lowest frequency of interest. Then, throughout a complete cycle, the maximum air particle velocity always occurs within the material. In practice, however, a gap may be left between the backing surface and the absorption material, as illustrated in figure 5.2a. Also, if the

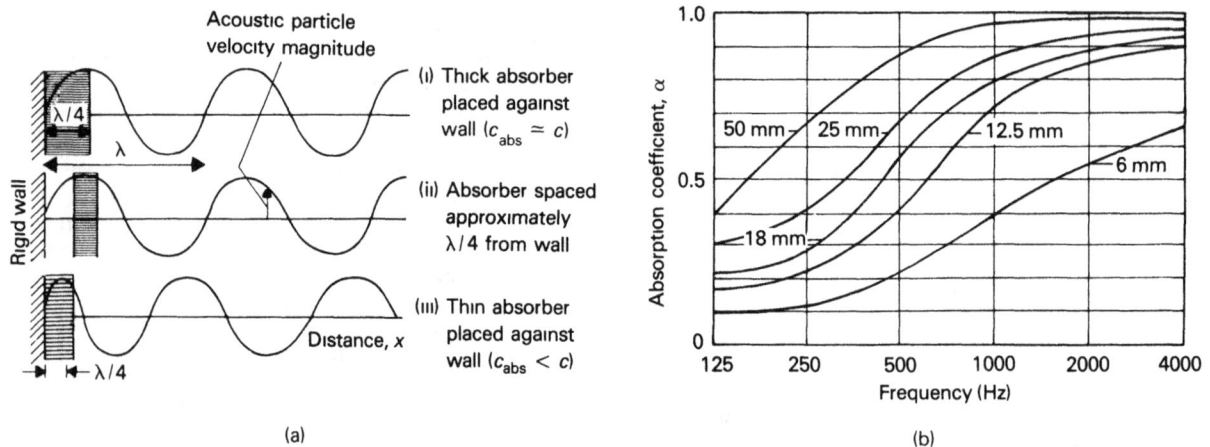

Figure 5.2 – (a) Choice of nominal positions of bulk absorber materials from a hard surface for optimum absorption, where c_{abs} is the velocity of sound propagation in the absorbent material and c is the ambient velocity of sound propagation. (b) Typical effect of material thickness on sound absorption (reference 6) (used with permission of Acoustical Publications).

speed of propagation through the absorber is substantially less than in free air, the effective wavelength is contracted and so the thickness of material needed is reduced. The influence of increased thickness on absorption, especially at low frequencies, is shown clearly in figure 5.2b.[6]

5.1.2 Description of Common Absorbers

The values of absorption coefficients depend on the physical mounting of the absorber, surface finish, and the immediate surroundings, as well as on their own innate characteristics such as thickness and porosity. Consequently, the values culled from tables should be regarded as approximate and may need to be changed depending on the situation and configuration used.

5.1.2.1 Foam

Foam is a very versatile material, especially because its acoustic properties are easily controlled and reproduced by appropriate modification of the manufacturing process. Polyester or polyether formulations are common, although the former permits a more close specification of cell size and reticulation. Cell sizes of between 5 and 50 pores per centimeter are readily achievable. Foam may be compressed under heat and pressure to a permanent set; moreover, it is easy to change the thickness and density. Furthermore, because foam represents a completely integrated material, no fiber shedding or material decomposition is likely to occur from the effects of sustained air flow or

vibration. Figure 5.3 shows some typical foams used for noise reduction.

5.1.2.2 Fibrous materials

Fibrous materials are usually constructed of glass fiber or mineral wool; their acoustical properties are varied by modifying the diameter of the fibers, packing density, and thickness of the finished product. Usually the fibers are cemented together at contact points with a suitable resin, although unbonded glass fiber blankets are also available. A major advantage of fiber blankets is that they have good inherent fire-resistant properties. However, they have to be used with great care in situations with high aerodynamic flows or vibration levels, since their delicate nature, especially at exposed edges, can lead to a great deal of fiber shedding.

To alleviate shedding, and to improve their attenuation characteristics, particularly in the lower part of the frequency range, a perforated impermeable outer cover is sometimes placed on top of the blanket. Figure 5.4 shows the typical acoustic effect of using such a protective covering on top of any porous layer. This essentially makes the material behave like a series of Helmholtz resonators (see section 5.1.2.3). The effect of the perforations is especially noticeable at low frequencies, when the total open area is less than around 25 percent. By varying the porosity, hole size, and hole spacing, a higher and increasingly sharper absorption peak may be obtained in contrast to the broadband absorption characteristics

46

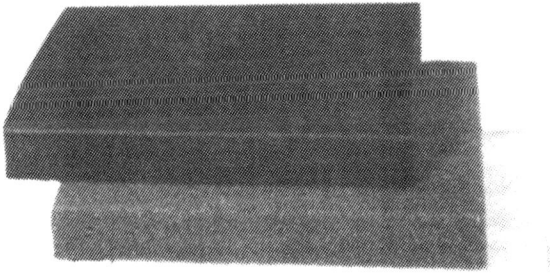

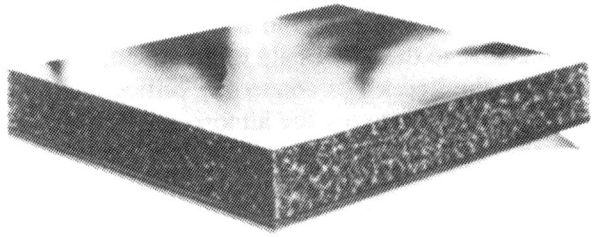

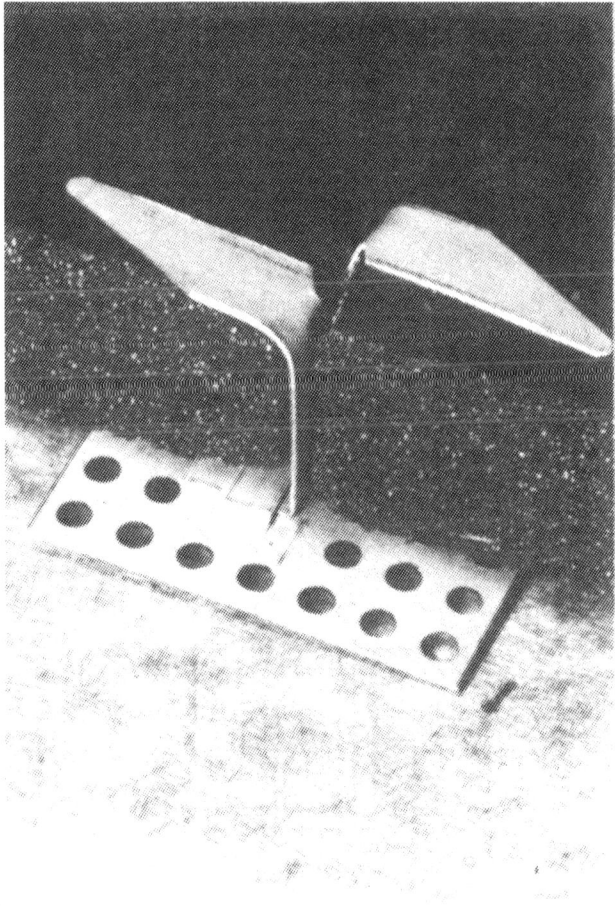

Figure 5.3 – Acoustical foams. Top, *by courtesy of Specialty Composites Corp.* Bottom left, *Eckousta-Foam being applied (Eckel Industries, Inc.).* Bottom right, *Eckoustic-Klip fastener for installing sound control materials (Eckel Industries, Inc.).*

Figure 5.4 – Effect of placing a perforated impermeable cover on a porous acoustical material.

common to most acoustical foams and fibrous materials. This must be traded off against the deterioration in high frequency performance because these components cannot diffract through the holes.

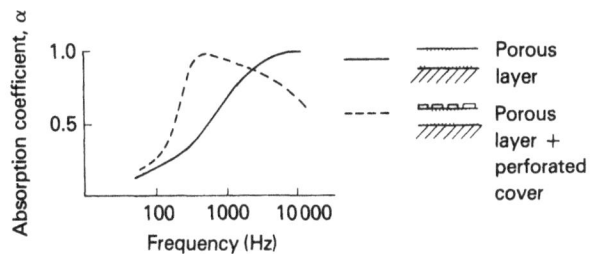

5.1.2.3 Helmholtz resonators

Helmholtz resonators are cavities connected to the ambient atmosphere by holes or slots in a well-defined geometrical pattern, as indicated in figure 5.5. When the dimensions of the cavity are small relative to the wavelength of the sound, then the air in the "neck" or connecting orifice oscillates like a plug between the atmosphere and the air in the cavity, resulting in high viscous losses. A very narrow absorption spectrum is associated with Helmholtz resonators, and they can be tuned for maximum absorption at a particular frequency by use of the following relationship[4]:

$$f_{res} = \frac{c}{2\pi} \left[\frac{S}{(L + 0.8 \sqrt{S})V} \right]^{1/2} \text{ Hz}$$

where

c is the speed of sound, m/s
S is the area of orifice, m^2
L is the length of orifice neck, m
V is the volume of cavity, m^3.

Normally, the height and width of the slots into the cavities are varied to achieve optimum absorption at a nominal frequency. Resonators are principally of value for attenuating prominent low frequencies below around 400 Hz. However, by filling the neck or cavity with a porous or fibrous absorber, it is possible to broaden the effective absorption range of these, albeit with some penalty in terms of the peak absorption, as shown in figure 5.5. The absorption range may also be broadened by a series of resonators coupled together laterally; a parallel configuration

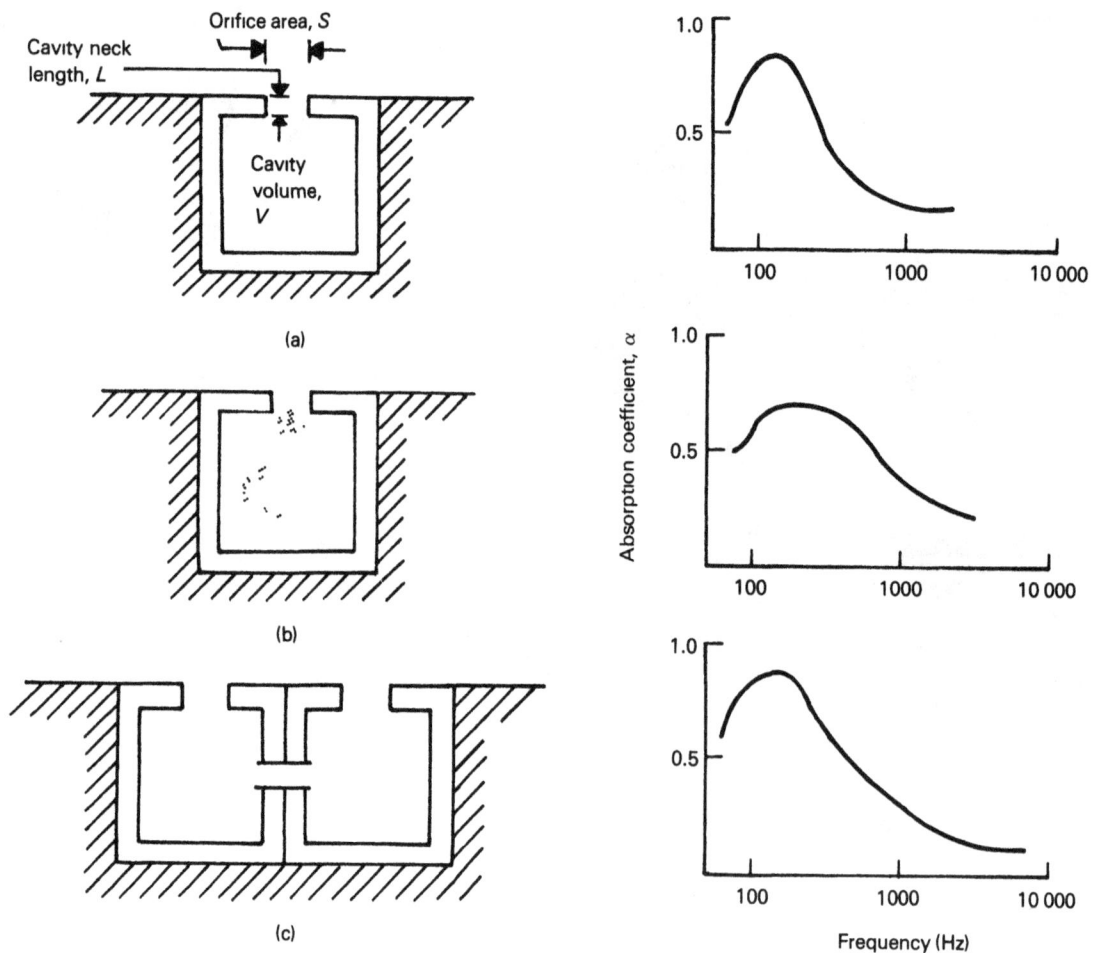

Figure 5.5 – Typical spectra for several Helmholtz resonator configurations. (a) Simple resonator. (b) Absorbent lined resonator. (c) Parallel-coupled resonator pair.

is particularly effective for sound entering at a tangent.[7] Reference 7 also gives details of optimum geometries for designing coupled resonators. An increase in energy attenuated of up to 50 percent as well as a significant increase in effective bandwidth may be gained. Figure 5.6 shows some examples of single resonator units used in industry.

Helmholtz resonators, mounted either as single units or as part of a series with complete access between cavities, such as perforated plate coverings (see section 5.1.2.2), are sometimes referred to as single degree of freedom systems because acoustic intensity at the resonator orifice can only propagate normally to the surface. It is also possible to stack several layers of coupled resonators, with different controlling parameters, on top of each other.

Extensive analytical work on the use of perforated plate liners in aircraft jet and fan noise reduction programs has highlighted several interesting phenomena of particular interest for duct noise control problems, where grazing incidence at the absorption liner boundary predominates. One study[8] showed that for the same porosity, flow resistance decreases as the number of holes in the plate is increased, with a concomitant decrease in hole size to retain the same total open area. It has also been shown that the nonlinear acoustic impedance effects at the absorber–air boundary, which is critical to the value of the absorption coefficient, may be treated as a high amplitude acoustic spectrum without grazing flow.[9]

The extreme environment of jet aircraft engines has resulted in considerable development of the broadband resistive resonator concept,[10,11] where some kind of metal fibrous layer is placed over a resonating absorber, often a metal honeycomb structure cut to vary cell depth across the panel. Such structures usually have a significantly broader attenuation bandwidth than the best perforate plate-faced systems; figure 5.7 compares the merits of the two systems. Figure 5.8 shows some honeycomb absorbers developed by NASA.

Devices for more general application based on similar principles have been developed[12]; although these are often more expensive than traditional absorbers, they are of use in rigorous industrial environments where, for example, high temperatures and contamination would detract considerably from the performance and suitability of foams and fibrous materials. Another design

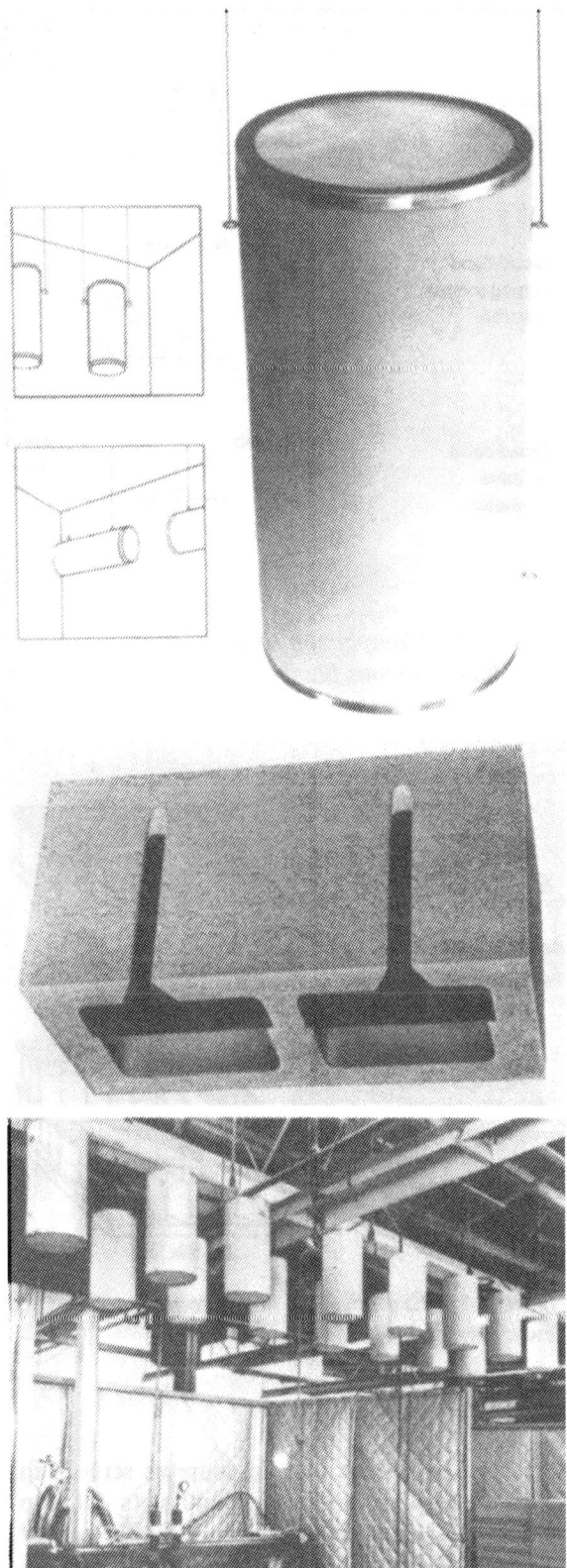

Figure 5.6 – Illustrations of Helmholtz resonator configurations (by courtesy of The Proudfoot Company, Inc.).

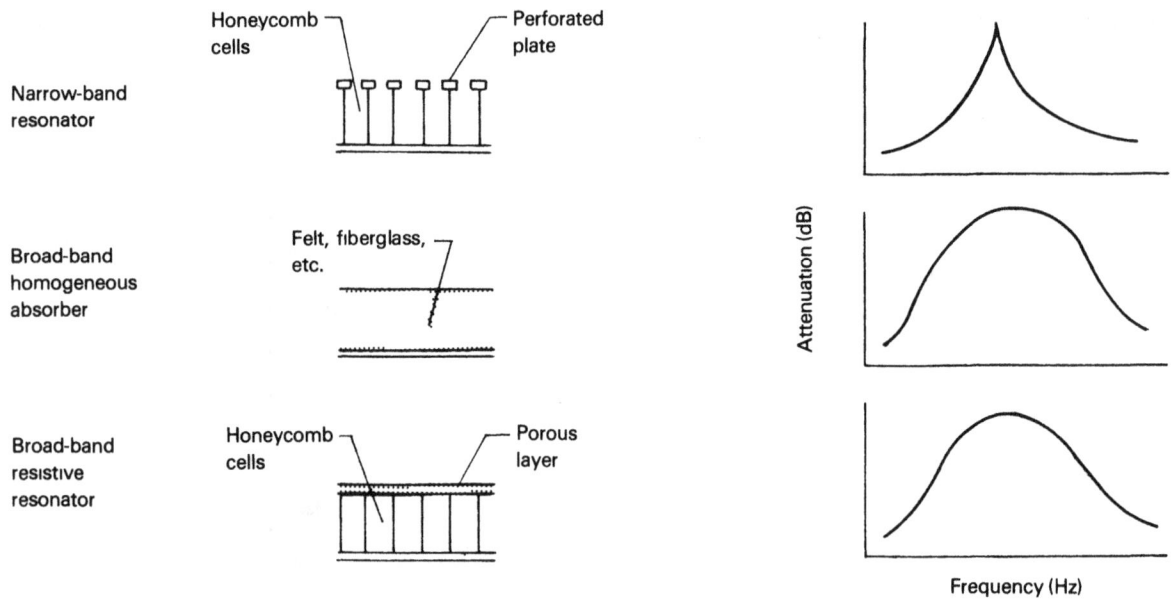

Figure 5.7 – Comparison of absorption spectra for perforated plate and broadband resistive resonators, and a homogeneous blanket absorber (reference 10).

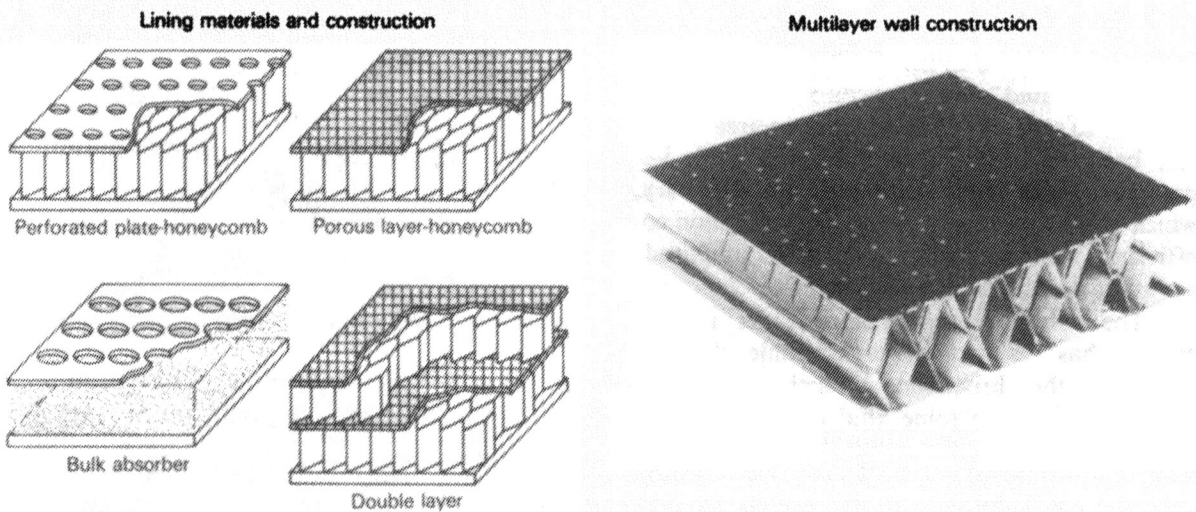

Figure 5.8 – Examples of honeycomb and perforate plate-faced absorber linings (by courtesy of NASA Lewis Research Center).

developed by NASA to help suppress screech in rocket motors utilizes longitudinal slots cut into mild steel liners with a backing cavity depth of around 2.3 cm; these slots essentially act as coupled Helmholtz resonators.[13] Combinations of slot widths were used to achieve the broadband performance needed in long and narrow combustion chambers of this type.

5.1.2.4 Functional absorbers

"Functional absorbers" is a generic term used for highly efficient sound absorbing panels suspended from a ceiling. They offer greater absorption than that possible with absorption on walls and are often used because of limitations on affixing panels directly to the room surfaces because of lighting, maintenance access requirements, and so on (see section 6.2.2). Their performance is determined not only by the intrinsic absorption properties of the material from which they are constructed, but also by their relative spacing and, to a lesser extent, their shape.

5.1.3 Applications of Sound Absorbers

Sound absorbers are especially useful in helping to control the reverberant sound field, although the degree of additional absorption needed to effect a reduction follows a logarithmic law of sharply diminishing returns (see section 6.2.2). Frequently, absorbers are used to line enclosures or partial barriers to prevent the particle build-up of reverberant sound caused by components reflected from the inner surfaces. This is illustrated in figure 5.9.

Resonators can be used successfully where prominent low frequency tones are present, such as electrical hum from main transformers or acoustic radiation from a machine with a low shaft rotational speed. They may be built into the structure of the building at the design stage, fitted as side-branch elements to ducts (see section 7.3.4), or added retroactively as independent units suspended in space (see figure 5.6).

Functional absorbers are used to control the reverberant sound field. Appreciable absorption can often be added to an industrial shop space only in the ceiling area, and the use of functional absorbers often conveniently avoids interfering with permanent fixtures. They are usually suspended vertically, and their individual performance improves substantially as the separation distance between them is increased.[14] Figure 5.10 shows some typical configurations and their associated sound absorption values. Staggering the panels horizontally and vertically can also enhance the total absorption.

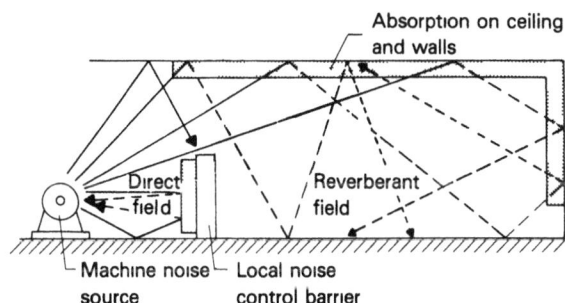

Figure 5.9 – Illustration of absorber use in helping to control direct and reverberant fields.

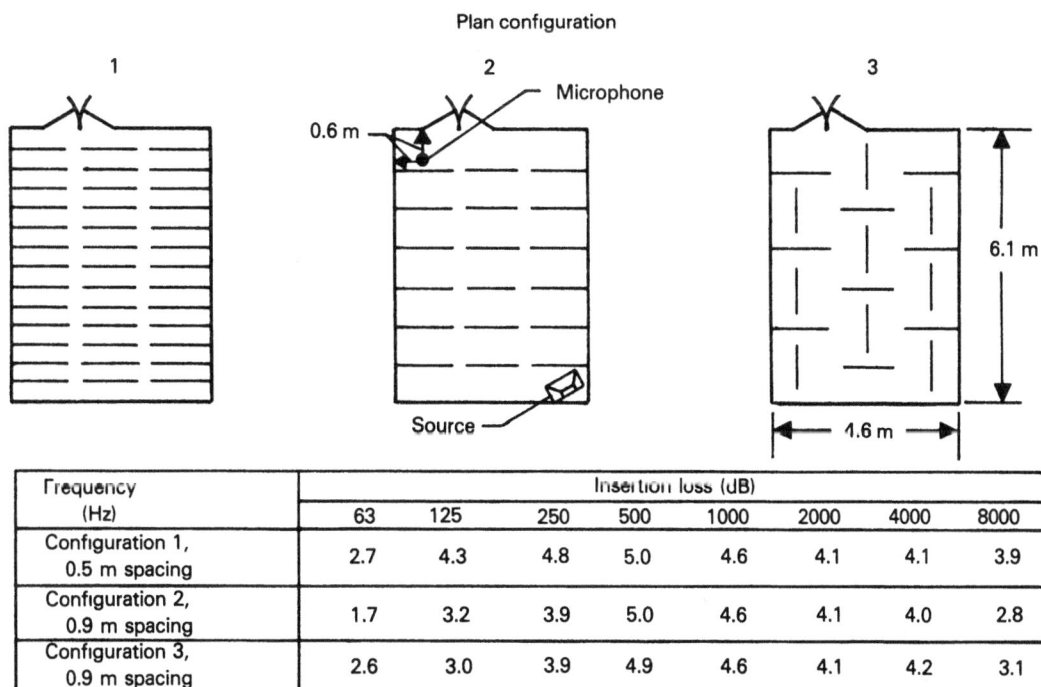

Frequency (Hz)	Insertion loss (dB)							
	63	125	250	500	1000	2000	4000	8000
Configuration 1, 0.5 m spacing	2.7	4.3	4.8	5.0	4.6	4.1	4.1	3.9
Configuration 2, 0.9 m spacing	1.7	3.2	3.9	5.0	4.6	4.1	4.0	2.8
Configuration 3, 0.9 m spacing	2.6	3.0	3.9	4.9	4.6	4.1	4.2	3.1

Figure 5.10 – Typical functional absorber arrays and associated sound absorption values (reference 14). Measurements made in reverberation room; panel dimensions are 1.2 × 0.6 × 0.04 m. (Used with the permission of the Noise Control Foundation and J. D. Moreland.)

5.2 INSULATING MATERIALS

5.2.1 Physical Theory

Insulating materials reduce the amount of sound transmitted from one space to another by maximizing the impedance mismatch. This mismatch reduces the amount of acoustic energy entering and leaving the wall or partition concerned. At the same time it is often useful to have a high internal damping factor (see section 5.3.1.1), which attenuates the acoustic energy traveling through the partition by converting it into heat. Consequently, many noise insulators are as massive as possible and quite limp, where stiffness is not a structural prerequisite.

The acoustical performance of any insulating material is described by a parameter known as the transmission loss, R, which is highly dependent on frequency. This is measured in decibels and is related to τ, the transmission coefficient defined in section 5.0, by

$$R = 10 \log (1/\tau) \quad \text{dB}$$

To compute the transmission loss it is necessary to make measurements in a specially designed facility.[15] This is performed by mounting a sample of the material under examination in an aperture between two reverberation rooms. The completely sealed aperture, and the massive construction of both rooms, ensures that the only significant path of transmission is through the test panel. Measurements are taken of the sound pressure level in both rooms, in sixteen one-third octave bands, after activation of a broadband noise source in one of them. The transmission loss is computed as

$$R = \text{NR} + 10 \log (S/A) \quad \text{dB}$$

where

NR is the SPL source minus SPL received in each one-third octave frequency band, dB

S is the total area of exposed test panel, m^2

A is the total absorption in receiving room, metric sabins.

In practical situations, possible discrepancies from this idealized measurement procedure may occur as a result of alternative paths of transmission and inhomogeneities in the separating partition; these are dealt with in section 5.2.3. A full range of typical sound insulation values is shown in table 5.2. Because sound insulation is only a relative quantity, a partition providing a net 30 dB of noise reduction reduces incident sound levels of 90 and 75 dB to 60 and 45 dB, respectively, at the other side.

5.2.1.1 Mass control

Figure 5.11 shows an idealized plot of transmission loss changes with frequency, illustrating the dominant features and controlling parameters. The most uniform portion of the curve is controlled solely by the mass of the partition; for a single panel the curve rises at the rate of 6 dB per doubling of mass and 6 dB per octave increase and hence is linear. In practice a number of expressions relate transmission loss to the frequency and to the density of the material. The normal-incidence mass law is an analytical expression derived by assuming a sound field only incident at angles normal to the surface, i.e., at 0°:

$$R_n = 20 \log fM - 48 \quad \text{dB}$$

where

R_n is the normal incidence transmission loss, dB

f is the frequency, Hz

M is the superficial density of insulator, kg/m^2.

Note: When dealing with insulating materials it is usual to speak of mass in terms of superficial density, which accounts for the total mass throughout a unit surface cross-sectional area of the material.

By making a further assumption that the sound field is diffuse and incident at all angles from 0° to 90°, then the random-incidence mass law is defined as

$$R_{\text{random}} = R_n - 10 \log 0.23 R_n \quad \text{dB}$$

where R_{random} is the random incidence transmission loss, dB.

The field-incidence mass law is a variant of this, again assuming a diffuse field, but only accounting for incident angles between 0° and 78°. In practice it has been found to be the most accurate and maintains a 6 dB increase with doubling of either mass or frequency, as contrasted to the 5 dB increase of the random-incidence law. It is given by

$$R_{\text{field}} = R_n - 5 \quad \text{dB}$$

where R_{field} is the field incidence transmission loss, dB.

Table 5.2 – Transmission loss values R for some typical noise control materials.

Material		Octave band center frequency (Hz)						
Description	Superficial density (kg/m²)	125	250	500	1000	2000	4000	8000
Lead								
1.5 mm thick	17	28	31	27	32	32	33	36
3 mm thick	34	30	32	33	38	44	33	38
Lead vinyl	2.5	11	12	15	20	26	32	37
	5	15	17	21	28	33	37	43
Aluminum								
20 gauge, stiffened	2.5	11	10	10	18	23	25	30
Steel								
22 gauge	6	8	14	20	23	26	27	35
20 gauge	7	8	14	20	26	32	38	40
18 gauge	10	13	20	24	29	33	39	44
16 gauge	13	14	21	27	32	37	43	42
Plywood								
7 mm thick	3	17	15	20	24	28	27	25
20 mm thick	10	24	22	27	28	25	27	35
Sheet metal								
Viscoelastic core	10	15	25	28	32	39	42	47
Plexiglass								
6 mm thick	7	16	17	22	28	33	35	35
13 mm thick	14	21	23	26	32	32	37	37
25 mm thick	28	25	28	32	32	34	46	46
Glass								
6 mm thick	15	11	24	28	32	27	35	39
9 mm thick	22.5	22	26	31	30	32	39	43
25 mm thick	62.5	27	31	30	33	43	48	53
Double glazing								
2 × 6-mm panels, 12-mm gap	30	23	24	24	27	28	30	36
2 × 6-mm panels, 188-mm gap	30	30	35	41	48	50	56	56
2 × 6-mm panels, 188-mm gap with absorbent in reveals	30	33	39	42	48	50	57	60
Double walls								
2 × 280-mm brick, 56-mm gap 2 × 12-mm plaster	380	30	38	47	61	75	80	81
2 × 12-mm insulating board 50 × 100-mm studs	19	16	22	28	38	50	52	55
Sandwich partitions								
2 × 5-mm plywood, 1.5-mm lead	25	25	30	34	38	42	44	47
2 × 18-gauge steel, 9-mm asbestos board	27	22	27	31	27	37	44	48
Masonry, brick, 125 mm thick								
Plastered both sides	240	36	37	40	46	54	57	59
Brick, 255 mm thick plastered both sides	480	41	45	48	56	65	69	72
Concrete block, 152 mm thick	176	33	34	35	38	46	52	55
Concrete, 103 mm thick	234	29	35	37	43	44	50	55

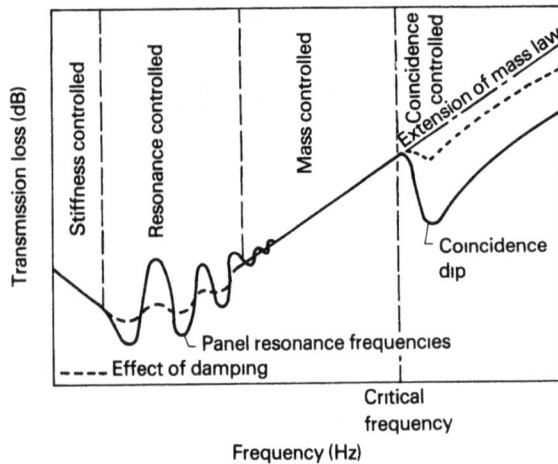

Figure 5.11 – Idealized transmission loss curve.

A rapid calculation of any of these quantities may be made using figure 5.12. All three relationships are valid only below the critical frequency of the panel of interest.

5.2.1.2 Coincidence effects

The critical frequency occurs when the speed of sound in air coincides with the propagation speed of free bending waves in the partition. Above this frequency, the free bending wave speed increases still further as a result of dispersive effects (see section 2.1.4), thus allowing for a range of frequencies at which the wavelength of airborne sound waves projected onto a panel coincides with the wavelength of free bending or flexural waves for that particular panel, as shown in figure 5.13. At the critical frequency this occurs for grazing-incidence airborne sound waves. At higher frequencies the effect occurs at angles increasingly less acute to the surface since the wavelength of flexural waves decreases less than that of corresponding airborne sound waves. The net result is a marked reduction in transmission loss because of easily induced vibration. This is shown in figure 5.11; the coincidence dip in the transmission loss curve usually extends over an octave or so above the critical frequency.

Table 5.3 lists critical frequencies for some typical insulating materials. The size of the coincidence dip in the transmission loss spectrum depends mainly on the homogeneity of the material and the presence of damping. Broad and shallow dips tend to occur with plastic and heavy masonry materials, while glass, most metals, and lining materials such as plasterboard and hardboards exhibit rather sharper and more pro-

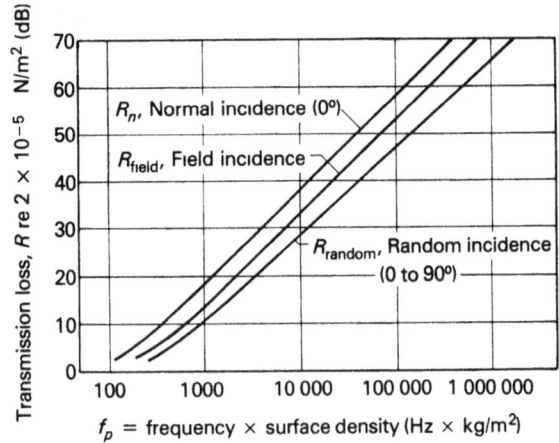

Figure 5.12 – Transmission loss curves valid for mass-controlled limp panels (reference 4). (From Leo L. Beranek, Noise and Vibration Control, Copyright © 1971 by McGraw-Hill, Inc. Used with the permission of the McGraw-Hill Book Company.)

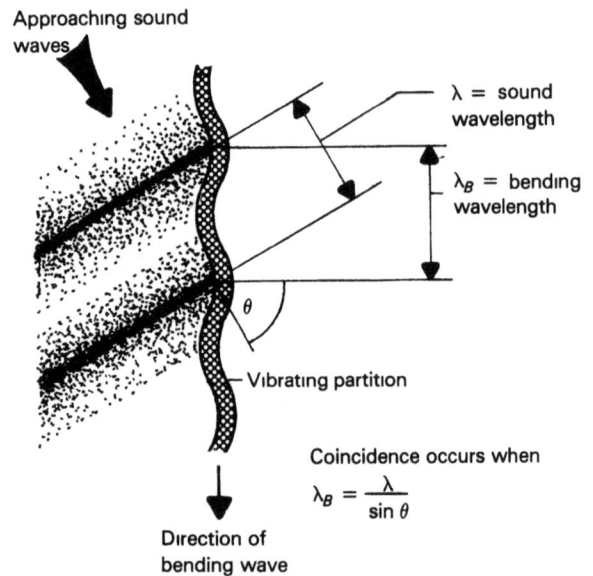

Figure 5.13 – Illustration of the coincidence effect. (From Ian Sharland, Woods Practical Guide to Noise Control, 1972; used with the permission of Woods Acoustics.)

nounced dips. In addition, denser materials tend to have a lower critical frequency. Consequently, coincidence effects may be mitigated by the use of additional damping, or by changing the thickness and hence superficial density of the panel, so that the dip occurs outside the frequency range of concern.

54

Table 5.3 – Critical frequencies and surface densities of some common materials.

Material	Critical frequency × surface density (Hz · kg m^{-2})	Surface density per unit thickness (kg m^{-2}mm^{-1})
Lead	600 000	11.2
Partition board	124 000	1.6
Steel	97 700	8.1
Reinforced concrete	44 000	2.3
Brick	42 000	1.9
Glass	39 000	2.5
Plexiglass	35 500	1.15
Asbestos cement	33 600	1.9
Aluminum	32 200	2.7
Hardboard	30 600	0.81
Plasterboard	32 000	0.75
Flaxboard	13 200	0.39
Plywood	13 000	0.58

From Ian Sharland, *Woods Practical Guide to Noise Control*, 1972. Used with the permission of Woods Acoustics.

5.2.1.3 Stiffness control

The dips in transmission loss at the low frequency end of the spectrum (see figure 5.11) result from the fundamental resonances in the panel itself. The position and amplitude of these resonances are governed by the inherent stiffness of the panels as well as by their inherent damping and size.

The effect of stiffening on these resonances has been studied by NASA and other organizations in the aerospace field,[16-19] to elucidate techniques for increasing the low frequency transmission loss of aircraft fuselage panels. Stiffening increases the fundamental resonance frequencies and essentially shifts the transmission loss curve to the right on figure 5.11, making it difficult for any major low frequency exciting frequencies to induce resonance conditions. However, increasing the density or net mass of the insulating panel conversely reduces the fundamental resonance frequencies. Thus achieving extra stiffness by increasing the panel thickness is not the most effective way to improve panel behavior in this area.[20] Ideally, stiffening or damping (discussed in section 5.3.1) should be incorporated in the areas where the largest panel deflections occur; these are normally the result of the first fundamental resonance mode of the panel.

Honeycomb lattices, fiberglass, and so-called "noiseless" or highly damped steel, all have excellent stiffness properties,[21] and the ultimate choice is determined by the necessity for other structural properties. Honeycomb materials are particularly valuable in that they possess a high stiffness/mass ratio; figure 5.14 shows the effect of applying a paper honeycomb onto an aluminum panel with a constraining layer on top. Stiffening with ribs, stringers, and so on increases the number of radiating areas operating independently. Such stiffening increases the acoustic radiation efficiency of the panel in the mass-controlled region, thereby reducing available insulation, while at the same time raising the fundamental resonance frequencies and improving the low frequency insulation performance. Figure 5.15 illustrates this result.

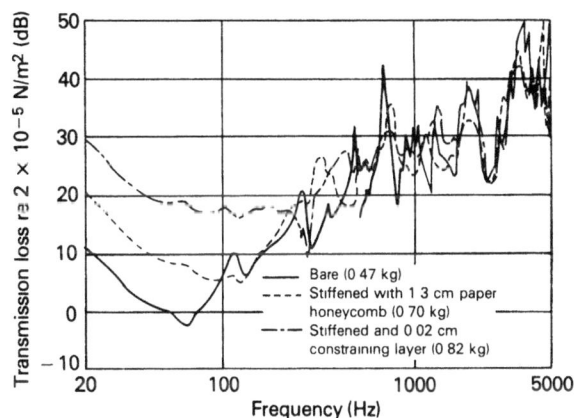

Figure 5.14 – Effect of paper honeycomb as a stiffener on an aluminum panel (reference 20). Panel dimensions are 45.7 × 45.7 × 0.1 cm; pure tone testing at normal incidence only to panel.

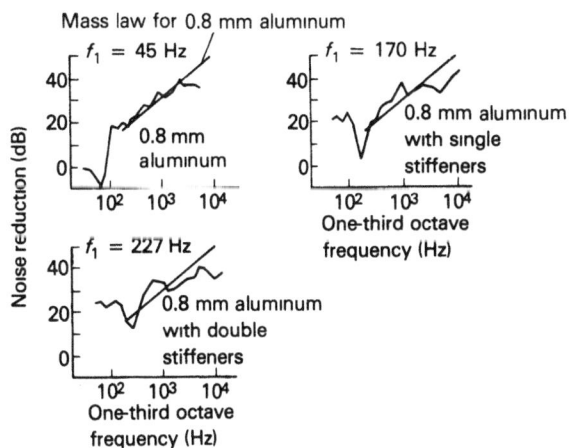

Figure 5.15 – Effect of rib stiffeners on an aluminum panel (reference 16). f_1 is the fundamental panel resonance.

The aircraft noise programs also establish that a closed cavity behind the insulating panel can be disadvantageous, because strong acoustic resonances may couple with the panel resonances to produce a very low transmission loss.[17,22] In fact, resonant modes in a receiving room are one of the principal causes of frequency-dependent variations in low frequency sound insulation,[23] and prediction methods have been developed for the overall response of a panel-room or panel-cavity system.[16,23] Structural geometry is of special concern here since it affects both acoustic and structural resonances. For instance, cylindrical and spherical panels are "membrane controlled," unlike flat rectangular or circular panels where "flexural control" is present.[24,25] The former are much more resistant to volume displacements, and design charts have been drawn for optimal geometries for panels with similar dimensions to the incident wavelength in the stiffness-controlled region. As a rough guide, "membrane control" is achieved when the length of a cylinder exceeds the geometric mean of the sum of the cylinder radius and wall thickness, and when the depth of a spherical segment exceeds the radius of curvature.[24]

Analytical work indicates that clamped-edge plates give better noise reduction below the fundamental frequencies,[17] although this is not true of the resonant frequencies themselves since an unclamped plate has higher inherent damping. A slope of -6 dB/octave increase normally prevails in the stiffness-controlled region, except where significant cavity-coupling effects occur, which may flatten the transmission curve somewhat.

5.2.2 Description and Rating of Common Sound Insulation Materials

As for absorption materials, it is often convenient to describe insulation in terms of a single number rating. This rating is known as the sound transmission class, STC, and is found by matching the transmission loss data against a standard transmission loss curve in a rigorously defined manner.[26] The STC rating is found from the value of the standard curve at 500 Hz. Basically, the measured values should be aligned such that they fall below the standard curve no more than 2 dB on average in each frequency band, and no greater than 8 dB in any one. An example of a typical calculation is shown in figure 5.16. STC ratings were developed specifically to

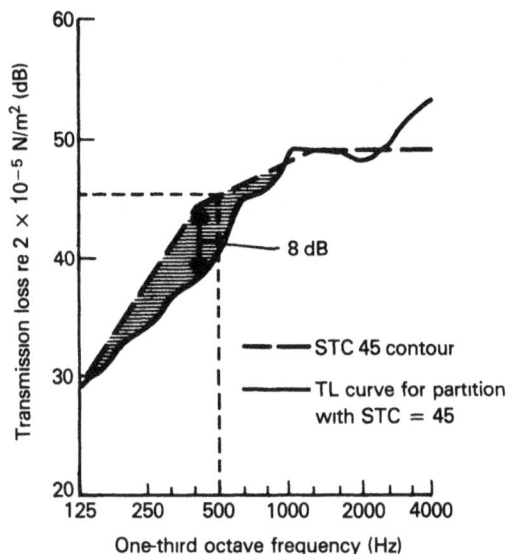

Figure 5.16 – Calculation of STC ratings.

handle noises with prominent spectra in the middle and high frequency ranges and, apart from obscuring serious dips in the transmission loss spectrum, they are not strictly relevant to many spectra found in industry, such as those with strong discrete tones or prominent low frequency components.

Generally, for good acoustical performance throughout the average industrial noise spectrum, a barrier should be as limp and massive as possible. It is also important that any weak paths, such as air gaps or poor insulators such as windows and doors, be reduced or eliminated since their effect is disproportionate to their total exposed area (see section 5.2.3). Hence for brick or lightweight walls with a relatively high porosity, it is advantageous to use a plaster or alternative impervious finish to keep the transmission loss as high as possible, even though this may not add substantially to the mass.

Lead is frequently used as an insulating material because of its high density, limpness, and good internal damping, which combine to produce a response which follows the mass-controlled transmission loss curve fairly uniformly without any serious coincidence or resonance dips (see figure 5.11). It may provide additional mass by including pulverized or sheet lead to form composite materials such as shot-loaded vinyls and other plastics. Sometimes these are fabricated in curtain form so that ease of access to a machine is maintained.

Where spatial restraints are not severe, double panels are a very convenient way of increasing the available insulation without adding to the overall mass. Since most transmission loss is obtained by impedance mismatching, subdividing the total weight into two separate partitions, thus increasing the number of air–insulator boundaries, gives considerable advantage, particularly in the mass-controlled region. To ensure maximum benefits, the gap between the two leaves should be as large as possible with at least one of the partitions resiliently mounted to prevent any structure-borne vibration from prejudicing the independent action of each panel. A comparison of double and single panels with the same total equivalent weight is shown in figure 5.17.[27]

Ideally, the slope of the double leaf transmission loss curve should be 12 dB/octave and the absolute value equal to the arithmetic sum of the two elements. However, in practice this does not occur for several reasons. At low frequencies a mass-air–mass resonance occurs when the air between the leaves acts as a spring and the whole system behaves as a single highly resonant unit. The frequency at which this occurs, f_{m-a-m}, is given by

$$f_{m-a-m} = \frac{600}{\sqrt{\dfrac{m_1 m_2 d}{m_1 + m_2}}} \quad \text{Hz}$$

where

m_1, m_2 are the superficial densities of two leaves respectively, kg/m^2

d is the separation distance, cm.

This effect reduces the overall performance of the double panel to below that for an equivalent single one unless the separation distance d is large enough. The adverse effects of standing waves in the air gap may be reduced by introducing absorption into the cavity; for double-glazing only the reveals may be treated. In addition, mechanical connections, if not resiliently isolated, can bridge the cavity and permit direct transfer of high frequency vibration across the structure. Staggering the studs to which the leaves are fastened is of value in preventing such transfer. Figure 5.18 illustrates several of these points.

Serious coincidence problems are avoided by making the two leaves of different materials or thicknesses. In one of the most comprehensive experimental and analytical studies of sound insulation elements to date,[28] a series of expressions for predicting transmission losses were derived. Most of these are for multiple-leaf partitions, which provide the most cost-effective means of gaining high insulation. Examples of double panels include double-glazing, gypsum board on stud

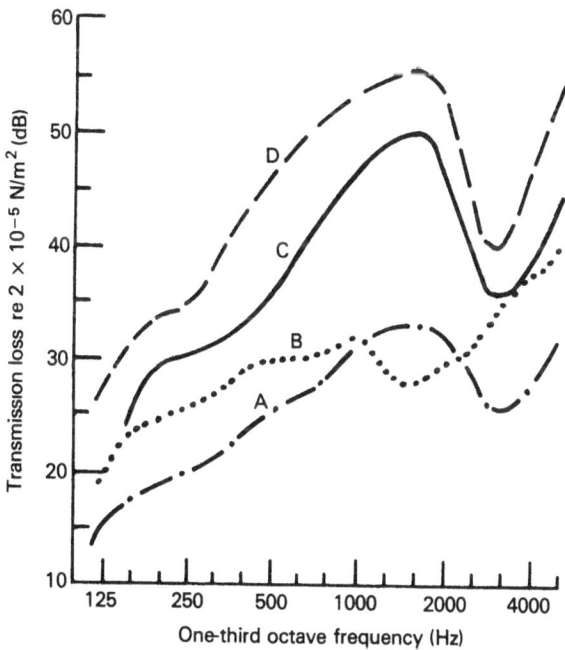

Figure 5.17 – Comparison of single and double leaf partitions of same equivalent weight (reference 27). A, curve for single-leaf 12-mm gypsum board; B, curve for two 12-mm leaves rigidly bonded; C, curve for two 12-mm leaves on either side of 90-mm steel channel studs; D, curve for configuration as in C, but with 50-mm fiber glass in stud space. (From Cyril M. Harris, Handbook of Noise Control, Second ed., Copyright © 1979 by McGraw-Hill, Inc. Used with the permission of the McGraw-Hill Book Company.)

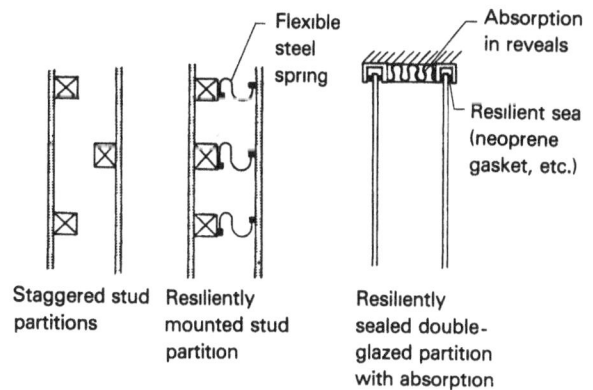

Figure 5.18 – Illustrative designs for double partitions.

walls, and wood- or expanded metal-lath ceiling and wall systems.

For noise transmission through a pipe or duct wall it is often best to add a second skin to the outside of the wall rather than to increase its mass or to add damping. Normally the outer skin is constructed of lead or cement screed, of around 10 to 20 kg/m^2 superficial density; a sufficiently thick resilient layer between the screed and the inner wall prevents undue coupling and allows the structure to act as a double-walled construction. This treatment must be applied over the entire length of a duct or pipe where high surface vibration exists, including side-branch turns, valves, and so on.

The most important characteristic in determining the overall transmission loss of a panel is its mass, which may be increased by increasing either the density or the thickness. However, high thickness, and hence stiffness inevitably reduces the coincidence frequency, often into an area of critical importance. In consequence, it would be desirable to have a material exhibiting high stiffness at low frequencies to provide insulation at frequencies below the panel fundamental, and a lower stiffness at high frequencies. One way of achieving this is by the use of laminated sandwich panels. One study[29] has considered several optimization procedures in terms of types of core and panel materials, or changes in the thicknesses of these, and concluded that two approaches may improve the overall insulation performance. One approach, which is sometimes impractical, uses less stiff and more massive panels effective over a greater frequency range; the other increases the core stiffness. At high frequencies, the shearing effect in the core reduces the total stiffness of the combination to something approaching that of the constituent materials.

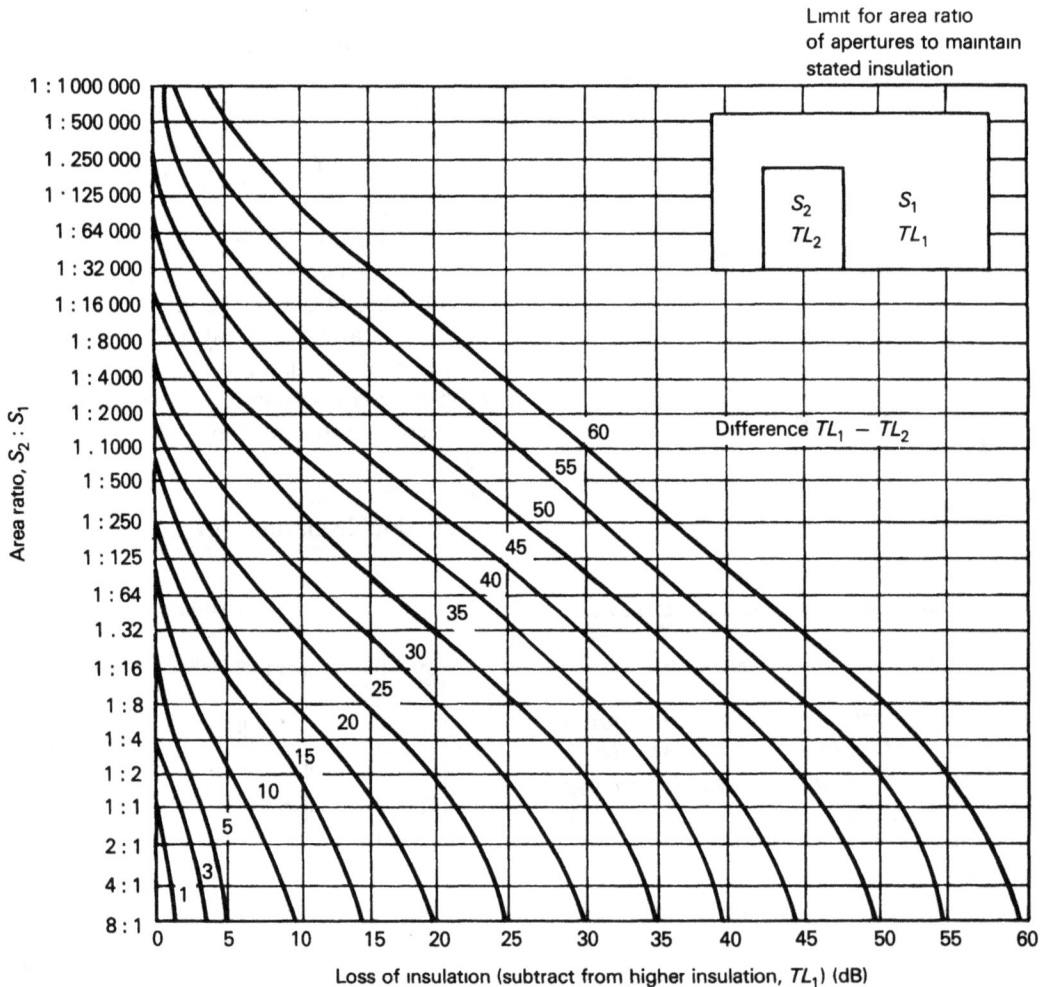

Figure 5.19 – Nomogram for computing the sound reduction index of composite partitions.

5.2.3 Installation of Sound Insulating Elements

It is imperative to make at least two calculations before actually installing sound insulation. The first calculation gives the effective transmission loss resulting from inhomogeneities within and around the installed panel such as air leaks, glazing, service conduits, and the like. The average transmission loss $\overline{R}$ is readily found according to the relations

$$\overline{\tau} = \frac{\tau_1 S_1 + \tau_2 S_2 + \cdots + \tau_n S_n}{S_T}$$

where

S_T is the total area of partition, m^2

S_n, τ_n are the area and transmission coefficient of the nth insulating element, respectively

$\overline{\tau}$ is the average transmission coefficient for the entire partition;

$$\overline{R} = 10 \log (1/\overline{\tau}) \quad \text{dB}$$

For convenience, a nomogram can be used for quickly calculating the overall transmission loss (figure 5.19). Two elements are taken at a time and their net transmission loss computed before incorporating a third and so on.

Example: For a wall with specifications shown as

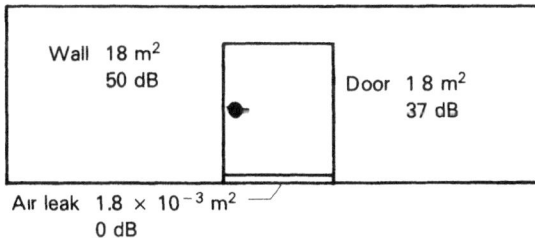

Net R for door + air leak:
    area ratio of door : air leak = 1:1000
Hence, from figure 5.19,
    high R − low R difference = 37 dB, loss of insulation = 8 dB
Therefore,
    net R, door + air leak = 29 dB
Net R for corrected door + wall:
    area ratio of door : wall = 1:10
Hence
    high R − low R difference = 50 − 29
                = 21 dB, loss of insulation = 9 dB
Therefore,
    net R, for entire partition = 41 dB

Note that if the air gap under the door threshold is ignored, an overall net transmission loss of 45 dB would be indicated, which shows the deleterious effect of small air leaks and other poor insulating elements on the overall performance. The calculation is performed across the entire frequency range, usually in octave bands.

The second calculation of importance involves an estimation of the actual change in sound pressure level at a point after the barrier or insulating element has been installed. This is known as the insertion loss; it should take account of the transmission loss, absorption of the reverberant sound field within enclosures, etc., and any flanking paths which may dominate the overall noise reduction after the insulation of the installed partition reaches a certain level. The use of absorption when constructing barriers or enclosures prevents the buildup of reverberant sound fields which would demand a higher insulation than initially planned. Examples of flanking transmission and the means of assessing and coping with it are discussed in section 6.2.3. Because the building and construction techniques used in the field may be quite different from those utilized when making laboratory measurements, it is often wise to make conservative performance predictions by subtracting 3–5 dB from the quoted transmission losses. Common causes of this degradation in performance are structural bridging caused by incorrect placement or construction of ties between walls, rubble in-filling, etc., or poor sealing of cracks around doors, windows, and conduit penetrations.

5.3 VIBRATION CONTROL MATERIALS

Vibration control materials can redirect or dissipate as heat acoustic energy entering a structure and appearing as structural vibrations. This is important, because once energy enters a solid material, it is possible for propagation to proceed along several different paths which may be difficult to predict, later to be re-radiated as sound waves in the atmosphere (see figure 6.13).

5.3.1 Vibration Damping

5.3.1.1 Physical principles of damping

Damping materials dissipate energy as heat somewhere within the structure to which they are applied, through friction, or less frequently, viscous effects, as well as the well-known hysteresis effect. Hysteresis losses occur because, after deformation caused by an applied force, many elastic materials do not regenerate the same amount of applied vibrational energy when returning to their original equilibrium position. Damping is typically most effective when conditions of large displacement amplitudes are found, and this is most common when resonance conditions prevail. For this reason they are usually used to complement insulator performance in the critical frequency and fundamental panel resonance ranges.

Damping materials may be used to control the radiation of sound from panels at the surface of a machine, where the resonant frequencies of the panels may be excited by either a broadband spectrum or discrete frequencies close to a resonant frequency. Damping also reduces impact noise. The addition of surface damping has a twofold effect in this regard: it considerably modifies the time history in which the largest frequency-transposed elements are well away from the lowest natural frequencies of the structure. Figure 5.20 shows how this may be achieved to produce a small force acting over a long time period as opposed to a large force acting for a short time.

Damping is measured in several different ways. The decay rate in decibels per second expresses the reduction in radiation from vibrating surfaces, once the exciting force is removed, with respect to a standard plate and shaking rig. Critical damping measures the damping necessary to just prevent oscillation and when expressed in the following ratio, ζ, is often used as a comparative measure between alternative materials:

$$\text{Damping ratio,} \quad \zeta = \frac{C}{C_c}$$

where

C is the damping coefficient of material under consideration

C_c is the critical damping coefficient.

A variant on this is the loss factor, η, which is related to the damping ratio by

$$\eta = \frac{1}{2\pi} \quad \overline{W_0}$$

$$\eta = \frac{2C}{C_c} \quad \text{or } 2\zeta$$

where

D_0 is the total energy dissipated by damping during one cycle

W_0 is the total vibrational energy of system.

Decay rates may range from 5 dB/s to over 80 dB/s and damping ratios from 0.5 to 20 percent of critical damping. Quite often, extra damping of significance can only be added when the damping ratio is less than 10%.

Until very recently there have been no really well-defined approaches to choosing damping materials for a particular situation. Successful application of damping control depends on a high degree of optimization after careful consideration of the following factors: temperature, frequency, static or dynamic loading, uncertainties in the precise nature of the incident spectrum, knowledge of the inherent damping of structures, and structural behavior under impulsive loading, i.e., linear or non-linear. However, owing in large part to pioneering work on acoustic fatigue and other vibration related problems in the U.S. aerospace industry, pioneered particularly by researchers at the Wright-Patterson Air Force Base, a number of important advances have been made.[30,31]

A temperature–frequency equivalence principle has been elucidated to permit the reduction of several highly specific parametric variations to a standardized and simple nomographic representation.[31] The performance specifications of several

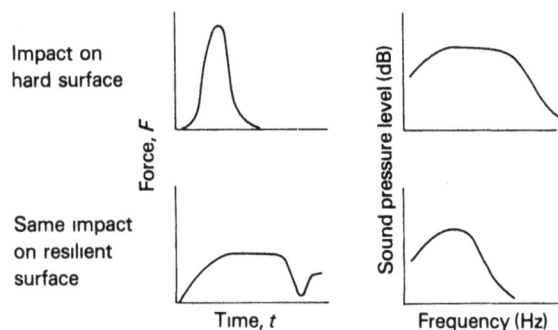

Figure 5.20 – Change in frequency spectrum for an impact due to damping. The area under the force-time curve in each case is the same and must always remain so when the total momentum involved in the interaction is constant.

dissimilar materials, from silicone elastomers to vitreous enamels, are adequately represented by this means. Figure 5.21 provides a typical example of this concept. The point of intersection between frequency and temperature is located, X, and the appropriate value of the reduced frequency, $f\alpha_T$, read on the lower abscissa scale at D. Knowing this value then allows the loss factor η and the modulus of elasticity E to be found from the curves plotted for the particular material concerned at B and A, respectively. Other design criteria for damping control are extant,[32,33] incorporating configurational changes for viscoelastic sandwich panels, and taking account of loss factors and structural resonant frequencies.

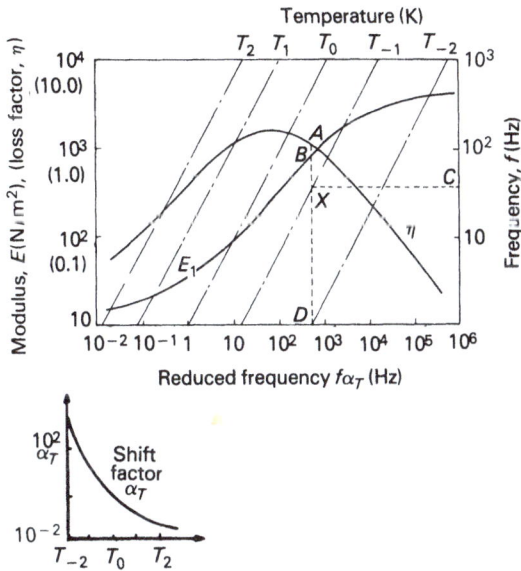

Figure 5.21 – Temperature–frequency nomogram for damping materials (reference 30). (Used with the permission of Bunhill Publications, Ltd. and D.I.G. Jones.)

5.3.1.2 Types of damping materials

Damping is effective only when intimate contact with the vibrating surface is obtained. To this end, secure bonding with an adhesive is essential. The two most popular types of damping are free layer and constrained layer. In general, effective free-layer materials are quite stiff and should be applied in thick layers, while constrained-layer materials are relatively pliable and often used in very thin layers. Free-layer materials normally can be trowelled, sprayed, or glued on to a surface, and coatings at least as thick as the vibrating metal surface or about 15–20 percent of its weight are often recommended. The operating principle of constrained-layer damping is that shearing action, as opposed to flexural or extensional motion, is considerably more efficient at damping vibrations. The constraining layer needs to be firmly attached to the viscoelastic core and important parameters to consider are as follows[32]:

(a) Base structure: thickness, inherent damping, frequency, density, and modulus of elasticity.

(b) Viscoelastic material: thickness, density, modulus of elasticity, and damping loss factor.

(c) Constraining layer: thickness, density, and modulus of elasticity.

The improvement in performance for a free layer of damping is demonstrated in figure 5.22, which shows that approximately five times the effectiveness is gained by doubling the thickness of the viscoelastic layer.

Analytical work[34] has shown that low frequency performance can be enhanced in viscoelastic sandwich panels by using a relatively hard core; this may provide up to 50 dB more reduction below 200 Hz. The operating range of

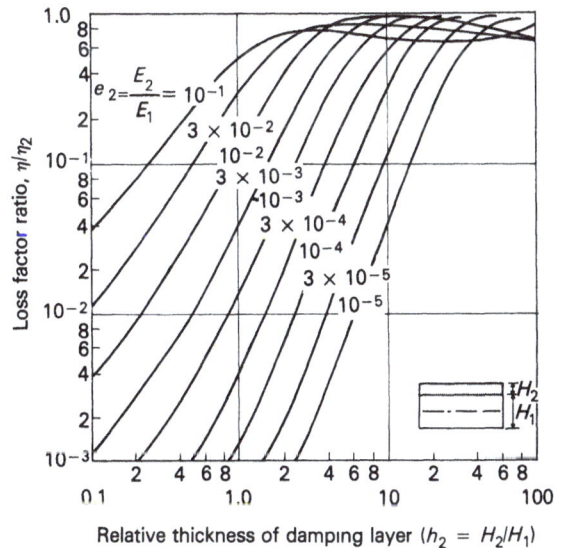

Figure 5.22 – Loss factor dependence on thickness for a free-layer damper (reference 4). n, loss factor of viscoelastic material; n^2, total loss factor of panel; E_1, E_2, elastic moduli of basic panel and viscoelastic layer, respectively. (From Leo L. Beranek, Noise and Vibration Control. *Copyright © 1971 by McGraw-Hill, Inc. Used with the permission of McGraw-Hill Book Company.)*

damping materials, in both temperature and frequency domains, may be improved by using multilayer treatments or by a judicious blend of damping compounds. It has recently been shown that some enamels have useful damping properties in the high temperature range 800° to 2000° F.[35] Tuned dampers operate within a relatively narrow range, being designed primarily to cope with just a few vibration modes well separated from any others. Figure 5.23 illustrates a few of these typical treatment configurations. Reference 36 includes a detailed discussion of optimization for each of these, including the effects of stiffness and geometrical characteristics. Table 5.4 shows the typical loss factors for a range of structural materials.

Much research and implementation remains to be done regarding damping control within the field of practical noise and vibration engineering. Moreover, in most extended metal structures such as air-handling ducts, a wide spectrum of resonant frequencies prevail, so that the effect of damping may be felt throughout a large frequency range.[36] Furthermore, in the mass-controlled region, increasing the mass certainly improves the transmission loss, but does not necessarily decrease the vibration amplitude in an equally unambiguous manner. Damping, on the other hand, affects both without entailing the fairly extensive modifications involved with stiffening or mass improvements.

5.3.2 Vibration Isolation

5.3.2.1 Physical principles of vibration isolation

Vibration isolators operate as low impedance materials inserted between high impedance elements in order to interrupt transmission paths. The mode of operation usually requires that a high proportion of the incident energy be absorbed and then returned before the end of each frequency cycle. Thus, even if the isolator materials are highly elastic, the vibrations cannot be transferred fast enough across it to produce effective transmission.

Notwithstanding the fundamental difference in the mode of operation and function between damping and vibration isolation, the two must be incorporated together for effective isolation, because large amplitude displacements occur when the loaded isolator exhibits resonant behavior at its fundamental frequency. Apart from amplifying the amount of transmitted vibration, hazar-

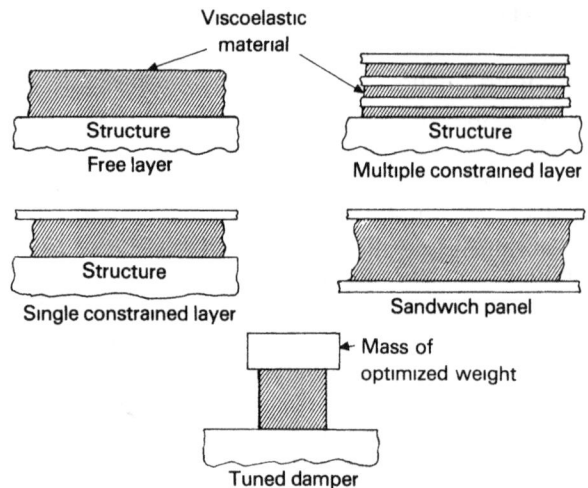

Figure 5.23 - Typical damping treatment configurations.

Table 5.4 - Typical loss factors of various structural materials.

Material	Loss factor, η
Aluminum	10^{-4}
Brass, bronze	10^{-3}
Brick	1 to 2×10^{-2}
Concrete	
Light	1.5×10^{-2}
Porous	1.5×10^{-2}
Dense	1 to 5×10^{-2}
Copper	2×10^{-3}
Cork	0.13 to 0.17
Glass	0.6 to 2×10^{-3}
Gypsum board	0.6 to 3×10^{-2}
Lead	0.5 to 2×10^{-3}
Magnesium	10^{-4}
Masonry blocks	5 to 7×10^{-3}
Oak, fir	0.8 to 1×10^{-2}
Plaster	5×10^{-3}
Plexiglass, lucite	2 to 4×10^{-2}
Plywood	1 to 1.3×10^{-2}
Sand, dry	0.6 to 0.12
Steel, iron	1 to 6×10^{-4}
Tin	2×10^{-3}
Wood fiberboard	1 to 3×10^{-2}
Zinc	3×10^{-4}

Notes: (1) Normal room temperatures.
      (2) Typical audio frequency range.

dous rocking movements may damage the mounted machine. Damping inhibits the degree of movement possible and concurrently reduces the transmission of vibration at this frequency. Conversely, too much damping limits the effective performance of the isolators, especially at higher frequencies. This is illustrated in figure 5.24, which shows a set of general curves for the transmissibility of a vibrating system. Transmissibility refers to the ease with which physical parameters such as force, velocity, or acceleration can be transferred across a system.

Isolators operate effectively as such only above $2f_1$ since below this value the isolator actually acts as an amplifier especially at the resonant frequency, f_1. The resonant frequency can be calculated by assessing the static deflection of the system involved when the isolator is under load, according to

$$f_1 = \frac{5}{\sqrt{\xi}}$$

where ξ is the static deflection, cm.

Static deflection usually changes linearly with loading for steel springs because the stiffness constant k in the following relationship remains constant:

$$f - \frac{1}{2\pi} \sqrt{\frac{k}{m}}$$

where
k is the spring stiffness, N/m
m is the mass loading, kg.

For many materials, however, stiffness changes nonlinearly with dynamic loading and it is therefore usually necessary to use the manufacturer's data to obtain the correct deflections and corresponding resonance frequencies.[37,38] The specifications on inherent damping provided in the material should also be closely examined.

As a general rule, 10 percent (0.1) transmissibility is a useful design criterion to meet. It is first necessary to determine the lowest forcing frequency likely to be encountered, which may be a shaft rotational frequency, for example. The amount of damping used rarely exceeds 10 percent, and after selection of this and an appropriate system resonant frequency, normally set at one-third or less of the value of the major lowest forcing frequency, it is possible to calculate the deflection which the chosen isolators must provide.

5.3.2.2 Types of vibration isolators

Solid, stiff pads of rubber may not be effective as isolators because their displacement is small and nonlinear relative to the load, making prediction of isolation performance difficult. Steel springs are more useful, especially for low frequency isolation, while high frequency performance may be improved considerably with the use of pads of elastomeric material at either end of the isolator. With regard to mounting, elastomers always perform better under shear loads, whereas cork, felt, plastics, and coils are best used in compression. Elastomeric isolators are fairly common in noise control; examples include natural, silicone, and butyl rubbers, as well as neoprene and combinations of all these. Their utility is extended because of their inherent damping and relatively low specific acoustic impedance. Reference 37 provides a good introduction to most isolators in common use and figure 5.25 depicts some typical units.

Pneumatic vibration-isolator systems offer several unique advantages relative to conventional isolators, which are discussed at length in reference 39. They are particularly good for insulating at very low frequencies and do not incur the same lateral instabilities as the lengthy steel

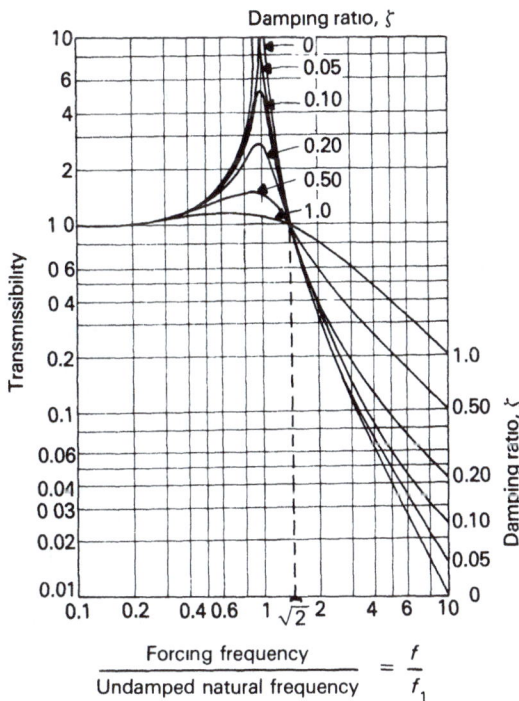

Figure 5.24 – Transmissibility of a vibrating system. f_1 is the undamped natural frequency of loaded system.

Figure 5.25 – Vibration isolation units (by courtesy of Vlier Engineering and Machinery Mountings, Inc.).

springs needed to perform the same task. The air pressure may be adjusted to support a higher load or to provide a constant level and height regardless of load changes. Typically they are rubber-bellows type units reinforced with nylon, with an internal pressure of around $7 \times 10^5 \, \text{N/m}^2$.

As a guide to the suitable selection of isolator mounts, generic types are grouped in table 5.5 with their appropriate ranges of normal static deflections.

Table 5.5 – Design ranges for types of vibration isolators.

Type	Static deflection (mm)
Steel springs	12.5 – 50
Rubber (in shear)	< 12.6
Cork, felt, highly-damped elastometers	< 6
Pneumatic isolators	No deflection requirements since internal pressure may be varied from atmospheric at $10^5 \, \text{N/m}^2$ up to $20 \times 10^5 \, \text{N/m}^2$

5.3.3 Application of Vibration Control Materials

As has already been intimated in section 5.3.1.1, damping can be particularly convenient as part of a noise control solution, because it is often easily applied as a free-layer treatment by troweling or spray techniques, and constrained-layer treatments are especially valuable in situations where spatial or structural restraints inhibit the addition of extra mass or changes in stiffness. Where impacting elements are involved, for example, a gravity chute used to channel products along a line, or drop-forge heads, or engine panels such as sump and valve head covers, either type of damping treatment may be applied. However, when some kind of elastomer is placed between the impacting surfaces involved, it is possible to modify the vibration spectrum imparted as well as to control the subsequent acoustic radiation.

Although the basic principles of isolator choice have already been presented, the methodology was based on the idealized situation of an infinitely heavy and rigid base mounting. In practice, this is seldom the case and unless the natural frequency of the isolator under load is about half that for the floor, then even low amplitude, low frequency forces will propagate through from the machine and may excite the fundamental frequency of the floor itself. As a rough guide, the static deflection of the isolator should be at least three to four times that of the floor, and approximately 90 percent isolation is produced when this is ten times the floor deflection. An estimate of the latter quantity is usually hard to make, but 1/300th of the total floor span is sometimes assumed. Figure 5.26 illustrates the corrective effects with the proper choice of isolator.

Another important consideration when choosing isolators is to ensure equal static deflection for each mount under the machine; otherwise, undue pitching and excitation of vibration modes higher than the simple vertical or axial modes may occur. This is achieved either by optimizing the stiffness for each separate mount, knowing the proportionate load, or by disposing the isolators equally about the center of gravity. The latter approach is often accomplished using an inertia block with a mass larger than that of the machine. This invariably results in a lowering of the complete unit center of gravity with respect to the isolators, which imparts extra stability. Figure 5.27 shows two examples of inertia block mounting. Pneumatic isolators are very flexible in this respect, in that they may be designed to accomplish maximum isolation and to maintain constant height regardless of the load conditions.[39]

Figure 5.27b highlights an important feature when isolating units, namely, the avoidance of bridging by a rigid mechanical connection between the machine and foundation. Often such short-circuits are overlooked in the process of designing isolators; commonly, service conduits, ducts, and hoses should be flexibly connected to surrounding structural surfaces.

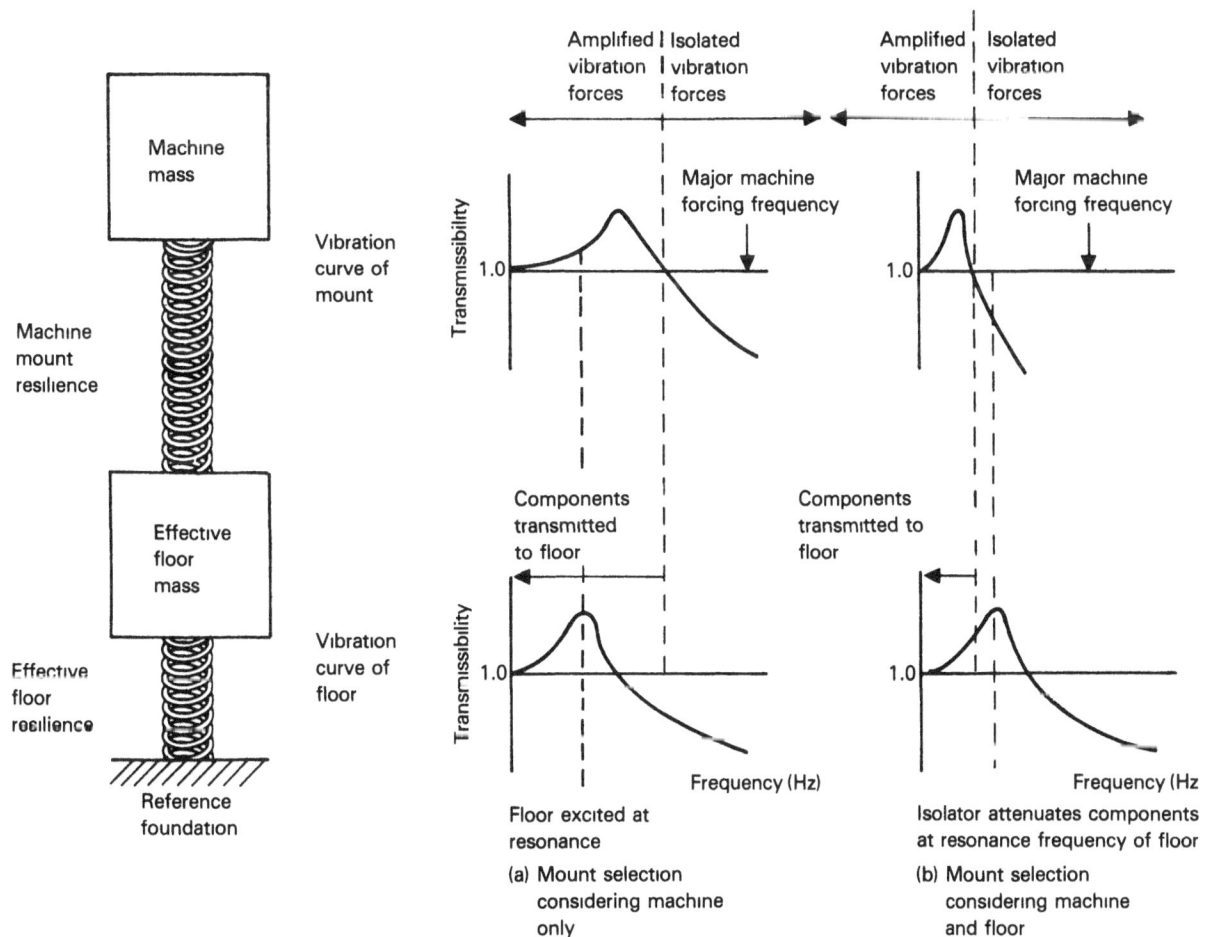

Figure 5.26 – Comparison of isolator performance for flexible and stiff floor mounts. (From Ian Sharland, Woods Practical Guide to Noise Control, *1972. Used with the permission of Woods Acoustics.)*

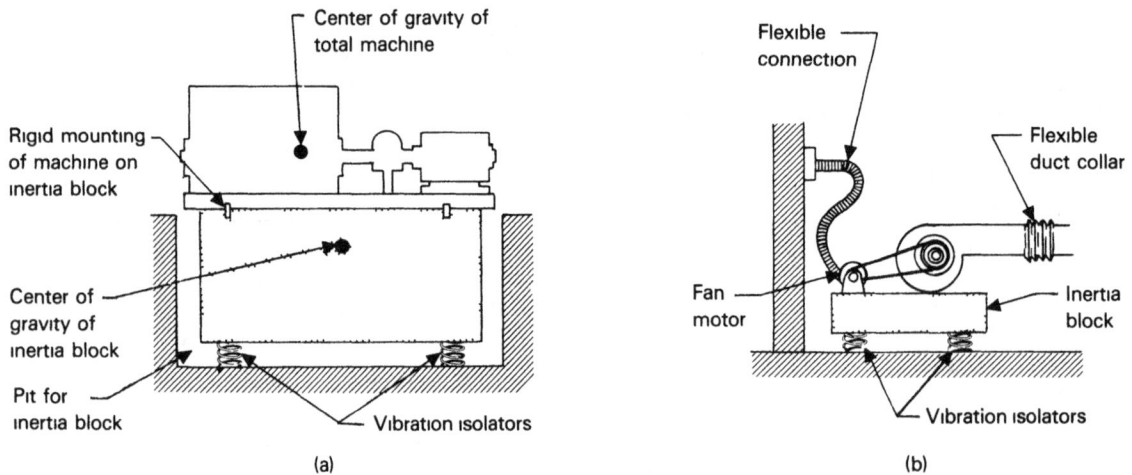

Figure 5.27 – Examples of various mounting installations for isolators.

Finally, it is important to ensure that lateral stiffness is no greater than vertical stiffness, in order to preclude serious sideways resonances. One way of trading off these two stiffness components is to orient the isolators so that their main axis is in direct alignment with the center of gravity of the machine, which may be some distance away. An excellent systematic approach to vibration isolation is presented in reference 40, which also contains formulas for calculating rocking mode resonances.

5.4 ENVIRONMENTAL AND LEGISLATIVE CONSIDERATIONS

The class of materials most sensitive to environmental constraints are traditionally the acoustic absorbers, especially those with a highly porous nature and consequent fine structure. Exposure to humidity, water spray, oil, grease, dirt, vibration, temperature extremes, and high fluid flow rates may all have a deleterious effect on acoustic performance and can influence the environmental suitability of the absorption material.

Protection against water and high humidity conditions can be achieved by placing a thin impervious plastic polyester sheet loosely over the face of the absorber or by incorporating a water-repellant treatment to stop waterlogging of the pores with concomitant loss of performance and increase in weight. Porous ceramics, open-cell polyurethane foams, and glass wool fibers all withstand humidity well.

The renewed interest of the aeronautics industry in evaluating the utility of incorporating bulk absorbers into engine duct linings has resulted in research examining the net effect of using water repellants. One such study[41] has shown that for a 1 percent decrease in flow resistance and a density increase of 3 percent after treatment with a repellant, some small changes in acoustic properties took place over the range 0.5–3.5 kHz. The attenuation constant, reflecting the absorption coefficient α, displayed no significant diminution in value except at the higher frequencies, where it dropped about 10 percent.

Machine enclosures often have absorption placed on the inside which brings the material into proximity with running equipment such as pumps, generators, saw blades, chucks, and the like. There is therefore a potential problem due to wicking of the fibers, etc., as a result of over-lubrication procedures, or to inadequately controlled air blow-off or cooling lubricant flows. The provision of good lubrication seals and deflecting shields may be of help; otherwise the absorbent may be modified by using imperforate thin sheeting as before. Normally it is only the high frequency end of the absorption spectrum which is slightly affected by these techniques, but to prevent further degradation in this, as well as to comply with sanitary and fire requirements, it is necessary to clean these surfaces regularly. This requires a strong surface, capable of withstanding repeated rubbing and cleaning with warm detergent solutions.[42,43]

For high temperature conditions, stainless steel or copper "metal wools" may be used as absorbers, and both are also of benefit where corrosive fumes exist.

In high fluid flow conditions, a retaining screen of perforated metal prevents fiber shedding from fragile materials, although it is important to secure the correct open-to-total-area ratio for suitable acoustical performance, as discussed in section 5.1.2.2. Even dense foams must have their raw edges sealed by thin plastic or a suitable paint surface.

The extensive use of honeycomb perforates in the extreme temperature and flow environment of jet engines has resulted in several optimization procedures for the acoustical performance of perforates, as discussed in the section dealing with ducts in chapter 7 (see section 7.3.2). One broadband sound absorber based on the honeycomb perforate principle[12] can be supplied with a special aluminum corrosion-resistant treatment. In addition to lending itself to ease of cleaning, it can be made self-draining, and possesses good thermal conductivity, which is of value where heat build-up near machinery housings can occur.

Barriers are usually constructed according to structural and acoustical requirements, but sometimes, when dealing with localized noise sources, it is necessary to retain some visibility of the work site. This can be accomplished by the use of shot-loaded transparent plastic barriers, for example, to shield conveyor systems. Although their performance may be diminished because of essential maintenance and access apertures, reductions of up to 10 dB(A) are achievable for a well-installed system. Flexible barriers made of loaded rubber or vinyl are useful where regular access between different parts of a shop floor is required, or for screening of small noise sources on a temporary basis.

Regulatory restrictions[44] have most impact within the food industry, where only glass, monel, or stainless steel are permitted direct contact with foods. Lead-bearing materials are forbidden, but barium can be used as a suitable alternative. There are also stringent cleaning requirements, which usually require that surfaces be resistant not only to high pressure steam and a certain amount of abrasion, but also to bacterial or fungal growth. Firebreak requirements demand a detailed consideration of ducting routes and the nonflammability of partitions and associated absorbing surfaces. Fiber dissipation may be a problem, not just from the point of view of acoustical degradation, but also because of the possible damaging effect of fibers on machine performance and their threat as a health hazard.

Finally, the manufacturer's recommendations regarding noise control material specifications and installation must be considered in implementing practical engineering measures to alleviate noise. Other factors to be considered include ease of maintenance after installation of noise control materials, and the effect on production rates both during and following installation of the designed measures.

REFERENCES

1. Kraft, R. E., R. E. Motsinger, W. H. Gauden, and J. F. Link, Analysis, Design, and Test of Acoustic Treatment in a Laboratory Inlet Duct, NASA CR 3161, 1979.
2. Hersh, A. S., and B. Walker, Acoustic Behavior of a Fibrous Bulk Material, AIAA Paper 79-0599, March 1979.
3. Lambert, R. F., Acoustical Properties of Highly Porous Fibrous Materials, NASA TM 80135, 1979.
4. Beranek, Leo L., ed., *Noise and Vibration Control*, McGraw-Hill Book Co., 1971.
5. ASTM, Standard Method of Test for Sound Absorption of Acoustical Materials in Reverberation Rooms, ASTM C423-66, 1966.
6. Purcell, W. E., Materials for Noise and Vibration Control, Sound and Vib., July 1976, p. 6.
7. Dean, L. W., Coupling of Helmholtz Resonators to Improve Acoustic Liners for Turbofan Engines at Low Frequency, NASA CR-134912, 1976.
8. Budoff, M., and W. E. Zorumski, Flow Resistance of Perforated Plates in Tangential Flow, NASA TM X-2361, 1971.
9. Rice, E. J., A Model for the Pressure Excitation Spectrum and Acoustic Impedance of Sound Absorbers in the Presence of Grazing Flow, AIAA Paper 73-995, October 1973.
10. Marsh, A. H., I. Elias, J. C. Hoehne, and R. L. Frasca, A Study of Turbofan-Engine Compressor-Noise-Suppression Techniques, NASA CR 1056, 1968.
11. Mangiarotty, R. A., Acoustic-Lining Concepts and Materials for Engine Ducts, J. Acoust. Soc. Amer., vol. 48, no. 3, 1969, p. 783.
12. Warnaka, G. E., N. D. Wonch, and J. M. Zalas, A New Broadband Sound Absorber, Internoise 76 Proceedings, 1976, p. 283.

13. Vincent, D. W., B. Phillips, and J. P. Wanhainen, Experimental Investigation of Acoustic Liners to Suppress Screech in Storable Propellant Rocket Motors, NASA TN D-4442, 1968.

14. Moreland, J. D., Controlling Industrial Noise by Room Boundary Absorption, Noise Con. Eng., vol. 7, no. 3, November–December 1976.

15. ASTM, Standard Recommended Practice for Laboratory Measurement of Airborne Sound Transmission of Building Partitions, ASTM E90-70, 1970.

16. Barton, C. K., Experimental Investigation on Sound Transmission Through Cavity-Backed Panels, NASA TM X-73939, 1977.

17. McDonald, W. B., R. Vaicaitis, and M. K. Myers, Noise Transmission through Plates into an Enclosure, NASA TP 1173, 1978.

18. Getline, G. L., Airframe Design for Reducing Cabin Noise, NASA Tech Brief, Summer 1978, p. 265.

19. Sengupta, G., Low Frequency Cabin Noise Reduction Based on the Intrinsic Structural Tuning Concept, NASA CR-145262, 1977.

20. Roskam, J., C. vanDam, D. Durenberger, and F. Grosveld, Some Sound Transmission Loss Characteristics of Typical General Aviation Structural Materials, AIAA Paper 78-1480, August 1978.

21. Roskam, J., F. Grosveld, and J. vanAken, Summary of Noise Reduction Characteristics of Typical General Aviation Materials, SAE Tech. Paper 790627, 1979.

22. Barton, C. K., and E. F. Daniels, Noise Transmission through Flat Rectangular Panels into a Closed Cavity, NASA Tech. Paper 1321, 1978.

23. Mulholland, K. A., and R. H. Lyon, Sound Insulation at Low Frequencies, J. Acoust. Soc. Amer., vol. 54, no. 4, 1973, p. 867.

24. Lyon, R. H., C. W. Dietrich, E. E. Ungar, R. W. Pyle, Jr., and R. E. Apfel, Low Frequency Noise Reduction of Spacecraft Structures, NASA CR 589, 1966.

25. Bailey, J. R., and F. D. Hart, Noise Reduction Shape Factors in the Low Frequency Range, NASA CR 1155, 1968.

26. ASTM, Tentative Classification for Determination of Sound Transmission Class, ASTM 413-70T, 1970.

27. Harris, Cyril M., ed., *Handbook of Noise Control*, Second ed., McGraw-Hill Book Co., 1979.

28. Wyle Laboratories, A Study of Techniques to Increase the Sound Insulation of Building Elements (prepared for HUD), available from NTIS as PB-222 829, 1973.

29. Lang, M. A., and C. L. Dym, Optimal Acoustic Design of Sandwich Panels, J. Acoust. Soc. Amer., vol. 57, no. 6, 1975, p. 1481.

30. Jones, D. I. G., Two Decades of Progress in Damping Technology, Air. Eng., Jan. 1979, p. 9.

31. Jones, D. I. G., Temperature-Frequency Dependence of Dynamic Properties of Damping Materials, J. Sound Vib., vol. 33, no. 4, 1974, p. 451.

32. Nashif, A. D., and W. G. Halvorsen, Design Evaluation of Layered Viscoelastic Damping Treatments, Sound and Vib., July 1978, p. 12.

33. Derby, T. F., and J. E. Ruzicka, Loss Factor and Resonant Frequency of Viscoelastic Shear-Damped Structural Composites, NASA CR-1269, 1969.

34. Vaicaitis, R., Noise Transmission by Viscoelastic Sandwich Panels, NASA TN D-8516, 1977.

35. Graves, G., et. al., On Tailoring a Family of Enamels for High Temperatures Vibration Control, Proc. 15th Annual Meeting, Society of Engineering Science (SES), Gainesville, FL, Dec. 1978.

36. Jones, D. I. G., Damping Treatments for Noise and Vibration Control, Sound and Vib., July 1972, p. 25.

37. Hochheiser, R. M., How to Select Vibration Isolators for OEM Machinery and Equipment, Sound and Vib., August 1974, p. 14.

38. Vér, I. L., Measurement of Dynamic Stiffness and Loss Factor of Plastic Mounts as a Function of Frequency and Static Load, Noise Con. Eng., vol. 3, no. 3, November-December 1974, p. 37.

39. Adair, R., The Design and Application of Pneumatic Vibration-Isolators, Sound and Vib., August 1974, p. 24.

40. Rivin, R. E., Vibration Isolation of Industrial Machinery—Basic Considerations, Sound and Vib., November 1978, p. 14.

41. Smith, C. D., and T. L. Parrott, An Experimental Study of the Effects of Water Repellant Treatment on the Acoustics Properties of Kevlar, NASA TM 78654, 1978.

42. Waggoner, S. A., J. F. Shackelford, F. F. Robbins, Jr., and T. H. Burkhardt, Materials

for Noise Reduction in Food Processing Environments, App. Ac., vol. 11, 1978, p. 1.

43. Miller, R. K., Acoustic Materials for the Food Process Industry, Noise-Con 73 Proceedings, 1973, p. 519.

44. Industrial Noise Services, Inc., Industrial Noise Control Manual (prepared for NIOSH), available from NTIS as DHEW Pub. No. (NIOSH) 75-183, 1975.

CHAPTER 6

Noise Control Procedures

Before implementing engineering or administrative measures to control noise, it is necessary to define the problem in regard to the persons affected by the exposure, the type and location of noise sources (see section 3.4.1), and the appropriate criteria for assessing the severity of the situation (see chapters 1 and 3 and the Appendix). Although every noise control problem must be examined individually, there are three separate components which should always be considered, namely the source, the propagation path(s), and the receiver(s). Acoustical treatment may be applied to any or all of these components with the general order of precedence being as listed; i.e., the most satisfactory solution usually results from noise reduction at or near the source.

There may be several different units or machines operating as sources, or perhaps a single one with several distinctly different components contributing to the noise field. In either case source identification and characterization are necessary to the proper design of control measures. For example, if a machine has several modes of operation, it is important to attribute sound pressure levels correctly to each distinct task. This may be accomplished, for example, by using high speed photography synchronized to the sound level measuring equipment[1] or some of the other techniques discussed in section 3.4.1.

Control of the source noise emission levels normally implies modifications to the structure, operational changes, or redesign of the physical configuration of the unit so as to relocate the noisiest components away from the most sensitive areas. Up to two percent of the energy input to a machine may be converted to acoustic energy and the particular characteristics of this may vary widely. An EPA report[2] lists noisy machines and typical levels for a wide variety of industries, and reference 3, which is essentially an examination of the spectral content of noise produced by several representative industries, dispels any misconception that there is a "typical" industrial noise environment. Approximately 11 percent of the large sample in the latter study was considered to contain pure tones.

Unfortunately, control at the noise source is not always practical for machinery already installed, and in such cases retrofit techniques must be applied at some stage along the path of propagation. These may involve partial or complete enclosure of the noise source, using a combination of insulation and absorption techniques; absorption at room boundaries to control the reverberant field; or vibration isolation where there is significant structure-borne transmission.

Alleviation of the noise field at the receiver is normally considered as a last resort or as a temporary expedient during the lead time required to implement some suitable engineering solution at the source or along the propagation path. Usually the most critical receiver is the operator of a machine or perhaps the occupant of the nearest dwelling where there is a community noise problem. The three alternatives are the provision of personal hearing protection, administrative control of the exposure duration, or complete enclosure of personnel in a well-insulated booth.

The wide variety of techniques which may be applied at either the source, path, or receiver should be evaluated in terms of cost, time needed to complete satisfactory noise control, availability of new components or quieter machines, and the effect on operations from the point of view of health and safety as well as productivity. The EPA has published a substantial survey of industrial noise[4] which includes source identification, in-plant levels, community noise impact, and application of suitable abatement technology.

The economic impact of noise control is not necessarily amenable to straightforward accounting, since there may be long-term benefits of improved productivity in a less stressful environment and better personnel relations and morale. However, as a rough guide, one study[5] defined the impact increment on production costs, Δ, as

$$\Delta = \frac{C_{nc}}{B \cdot Y}$$

where

C_{nc} is the cost of noise control
Y is the life of machines, years
B is the annual productivity.

This study essentially examined retrofitting of quieter components to a specific type of machine, but a more general discussion of the economics of engineering design for noise control may be found in reference 6. Net costs may be reduced somewhat by eliminating any hearing conservation program if the control measures reduce exposure conditions to below the relevant standard.

This chapter discusses each part of the total noise problem in some detail, and a detailed discussion of some common noise sources is given in chapter 7.

6.1 NOISE REDUCTION AT SOURCE

Some typical steps which may be taken to achieve source noise reduction include substituting a different mode of operation such as welding for riveting; using entirely new, quieter equipment which may also have a direct production benefit; and modifications to existing equipment such as balancing rotating parts, improving gears, using higher-powered motors with a consequent lower speed, and relocating service components of a large machine, such as generators and fans. The major nonacoustic considerations which enter into the final choice are the cost of alternative noise reduction schemes which achieve the same goal and productivity constraints due to post-control restricted access or machinery speed reductions. The two general categories of noise source are vibratory and aerodynamic; table 6.1 shows some noise level ranges for a variety of these sources.

6.1.1 Vibratory and Impact Sources

Vibration usually results from the excitation of resonances in machine components by cyclic or impulsive forces. Vibration may be caused by eccentric rotation, rapid changes in translational or rotational velocity, or impact on other machine parts. The most fundamental solution is to reduce the amplitude of vibration and hence the energy available for acoustical radiation. Reducing the vibrating area attenuates the total sound power radiated, and changing the dominant frequency of vibration can alleviate the forcing conditions on the structure or move the spectrum to a region that is less sensitive with respect to the human ear.

Amplitude reduction can be accomplished by keeping all contact surfaces smooth and well lubricated, which requires a thorough and regular maintenance schedule. Where possible, all rotating parts should be dynamically balanced, and their mass reduced as much as possible. This is particularly important for fans. Radiation from machine panels may be controlled by vibration isolation of the particular component. It is normally possible to detect the crucial radiating components by making accelerometer measurements and considering the panel area involved. Vibration isolation may be achieved as described in section 5.3.2. The acoustic power PWL radiated from a panel surface may be calculated as[7]

$$PWL = \sigma <\overline{v^2}> \varrho cS \quad W$$

where

σ is the radiation efficiency of the panel, usually approximately 1
$<\overline{v^2}>$ is the space-averaged mean square velocity of panel surface, m/s
ϱc is the specific acoustic impedance of air, $N \cdot sec/m^3$
S is the radiating area of panel, m^2.

Impact noise can be suitably reduced with the addition of some form of damping in order to apply the force more gradually and continuously (see section 5.3.1.1 and figure 5.20). This may also be achieved by canting blades and edges on die cutters, punch presses, and stampers. For blade cutting operations, increasing the number of blades reduces the impact load on each one, and shifts the generated spectrum toward higher frequencies where the noise field is often easier to control.

A good example of source noise control, furnished from an extensive series of studies on woodworking noise conducted at North Carolina

State University[8-10] and for which partial support came from NASA, is the redesign of wood planers.

Table 6.1 – Noise levels at the operator position for a range of industrial machinery and processes (reference 4).

	Noise levels (dB(A))								
	80	85	90	95	100	105	110	115	120
Pneumatic power tools (grinders, chippers, etc.)			←———————————————→						
Molding machines (I.S., blow molding, etc.)					←——→				
Air blow-down devices (painting, cleaning, etc.)			←————————→						
Blowers (forced, induced, fan, etc.)		←——————————→							
Air compressor (reciprocating, centrifugal)			←———————→						
Metal forming (punch, shearing, etc.)		←————————→							
Combustion (furnaces, flare stacks) 6.1 m (20 ft)		←————————→							
Turbogenerators (steam 1.8 m (6 ft))			←→						
Pumps (water, hydraulic, etc.)		←——————→							
Industrial trucks (LP gas)			←→						
Transformers		←→							

State University[8-10] and for which partial support came from NASA, is the redesign of wood planers. The most interesting innovation in this case was the successful deployment of a helical blade cutterhead,[8] which introduces point impact and reduces force fluctuations on the work piece (see figure 6.1).

Damping solutions are particularly appropriate when a large number of resonances are excited. An example of damping control is furnished in figure 6.2, where a vibratory parts feeder used to orientate work pieces was quieted. The noise sources were chiefly the impact of the parts with each other and with the bowl itself. An acoustical hood was also placed over the bowl to inhibit airborne radiation.

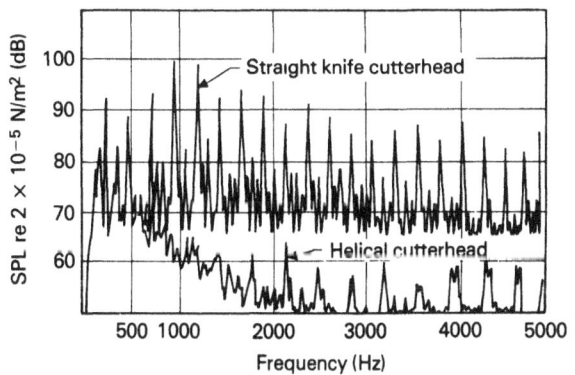

Figure 6.1 – Comparison of sound spectra for helical and straight knife cutterheads (reference 8). Board width, 30.5 cm; thickness, 2.54 cm; cut depth, 0.32 cm. (Used with the permission of the Noise Control Foundation and J. S. Stewart.)

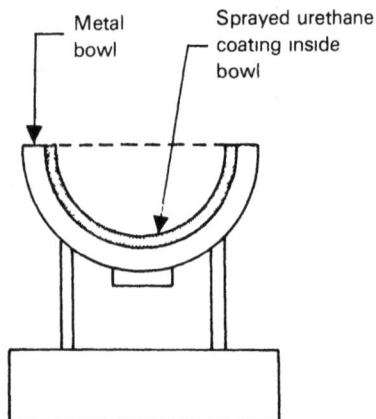

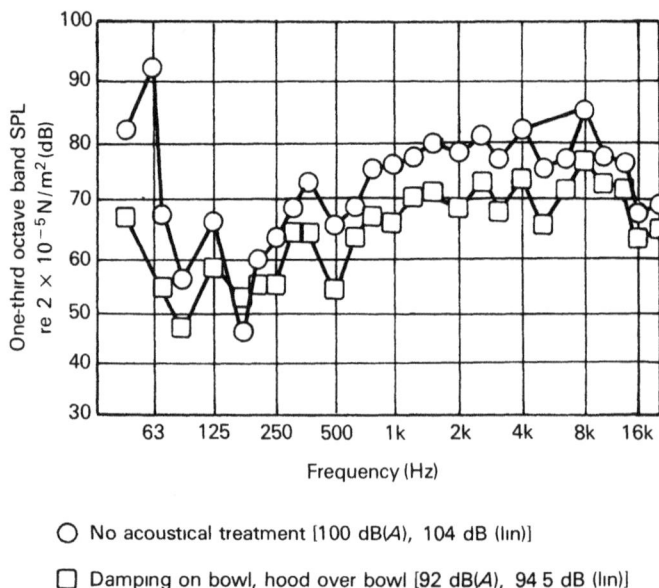

Figure 6.2 – Acoustical treatment of vibratory parts feeder and attendant reduction in noise spectrum (reference 11).

6.1.2 Aerodynamic Sources

Aerodynamic sources usually generate broadband noise because of turbulent, vortical motions in gases. This category includes the noise produced by flames, jets, valves, fans, and relatively high speed fluid flows. The noise level is usually related to the turbulence intensity, which in turn is a function of the velocity of the medium concerned. Thus it is often possible to make substantial reductions in the noise by increasing cross-sectional areas and hence decreasing net flow velocities, and by eliminating right-angled bends or constrictions in the ducting. There is a whole gamut of silencers which can be deployed to attenuate air jets and exhausts venting to atmosphere; these are dealt with in section 7.1.2. Other subsections in section 7.1 deal with valve, combustion, and fluid flow problems, while section 7.2.2 is concerned with fan noise reduction techniques.

6.2 PROPAGATION PATH AND REVERBERATION FIELD CONTROL

There are few models, other than highly idealized mathematical models, that describe the behavior of sound propagating through complex industrial environments. However, an attempt has been made to model a complex noise field by con-sidering it to be composed of cells with boundaries, appropriately defined as building walls, machinery, pipe arrays, etc.[12] Statistical energy analysis, SEA, is being increasingly used to investigate sound propagation through extended, complex structures. The uses and possible applications (see section 6.2.4) of SEA are presented here because the technique helps to solve complex acoustic propagation problems through aircraft and space vehicle fuselages.

The following sections describe specific mathematical relationships and design criteria to be used when considering control of structure-borne or air-borne sound between the source and receiver positions.

6.2.1 Acoustic Insulation Using Barriers and Enclosures

The basic principle involved in insulating against acoustic energy propagation is to provide impedance mismatching, as discussed in section 5.2.1. The most common forms of construction used are close-in acoustic hoods, solid barriers, and complete enclosures. These are most effective when used to limit the direct or free field of the source, and thus it is important to ensure their placement within this area. In order to achieve this, the following series of relationships may be used to predict the total sound pressure level at a particular point resulting from contributions from both the direct and reverberant field.

From section 3.4.4, the direct field intensity is governed by

$$\frac{p_\theta^2}{\varrho c} = \frac{WQ}{4\pi r^2}$$

where

ϱc is the characteristic acoustic impedance, $\mathrm{N\cdot sec/m^3}$
p_θ is the sound pressure at angle θ, $\mathrm{N/m^2}$
Q is the directivity factor at angle θ
r is the distance from source, m
W is the source acoustic power, W.

The reverberant field intensity is governed by

$$\frac{p^2}{\varrho c} = \frac{4W}{R_c}$$

where

R_c is the room constant given by $S\bar{\alpha}/1-\bar{\alpha}$, $\mathrm{m^2}$
S is the area of room with average absorption coefficient α, $\mathrm{m^2}$.

Combining these,

$$\frac{p_\theta^2}{pc} = W\left[\frac{Q}{4\pi r^2} + \frac{4}{R_c}\right]$$

In logarithmic form, using the intensity level–sound pressure level equivalence principle,

$$\mathrm{SPL}_\theta = \mathrm{PWL} + 10\log\left[\frac{Q}{4\pi r^2} + \frac{4}{R_c}\right]$$

The leveling off in the decay rate from 6 dB per doubling of distance when traversing from the direct field into the reverberant field is governed by the relative magnitude of r and R_c, and the transition point r_c, between the two fields is shown in figure 3.11 (see also section 6.2.2). The above relationship applies only where a spherical radiation pattern may be assumed as from a point source (see section 3.4.3).

For transmission across a solid boundary, the following relationships apply:

Room to room transmission:

$$\mathrm{SPL}_2 = \mathrm{SPL}_1 - R + 10\log S_T - 10\log A \ \ \mathrm{dB}$$

where
SPL_1, SPL_2 are the incident and received sound pressure levels respectively, dB
A is the total absorption in receiving room ($= S\bar{\alpha}$, metric Sabins)
S_T is the total area of common partition between rooms, $\mathrm{m^2}$
R is the composite transmission loss of partition, dB.

Notes:

(1) $A \approx R_c$ when $\bar{\alpha} < 0.3$.
(2) Reverberant field excitation only is assumed; the machine or source should be located as far as possible from the common room boundary to prevent direct field excitation.
(3) Flanking paths via the floor, wall, or ceiling structures may be present, thus significantly increasing the received level, SPL_2.

Room to free space transmission:

$$\mathrm{SPL}_2 = \mathrm{SPL}_1 - R - 6 \ \ \mathrm{dB}$$
(Close to wall)

where
SPL_1, SPL_2 are the incident level and level at external wall surface, respectively;

$$\mathrm{SPL}_2 = \mathrm{SPL}_1 - R + 10\log S - 20\log r + \mathrm{DI}$$
(At a distance from wall)

where
r is the distance from wall, m
S is the area of transmitting surface, $\mathrm{m^2}$
DI is the directivity index as defined in section 3.4.2.

Notes:

(1) Free-field conditions external to the boundary are assumed.
(2) Shape and proportions of wall determine its directivity characteristics.

75

6.2.1.1 Barriers

Barriers generally offer little noise reduction potential except in the acoustic shadow behind the barrier, where mid to high frequency sound cannot easily diffract over the top or around the barrier edges. The placement of the barrier close to the source or receiver is important, as this increases the path length difference defined in figure 6.3, which is the most critical parameter in determining its performance. Normally, a broadband sound spectrum is assumed in making barrier calculations, but if prominent pure tones are present and the source or receiver are some distance from the ground, up to three additional propagation routes may be created; these routes can exhibit significant destructive or constructive interference.

Barriers are usually floor standing and should be wider than they are high. They need not be particularly massive, since 20 dB(A) is about the maximum net attenuation attainable in practice and the attenuation for propagation directly through the barrier itself need not be more than 10 dB above the overall attenuation desired.[13] The same criteria apply to acoustic hoods used to shield only a relatively small area of a machine, except that their smaller dimensions tend to make them worthwhile only when dealing with high frequency sounds such as air ejectors. Normally they are used in conjunction with well-placed absorbers, as indicated in figure 5.9, to prevent flanking circuits resulting from reflected, and ultimately reverberant, sound build-up at the receiver.

According to the Fresnel theory of diffraction, the performance of barriers may be predicted as a function of the value of the Fresnel number, N, calculated as shown in figure 6.3.

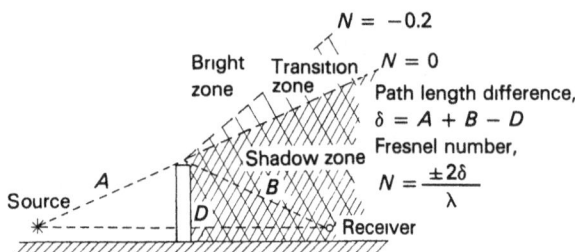

Figure 6.3 – Barrier Fresnel zones and calculation of path length differences. Sign for N *chosen plus or minus, depending whether receiver is in the shadow or bright zone, respectively.*

The major limitation on barrier attenuation results from the finite size of the barrier. This may be overcome to some extent by making use of any local features such as storage bins or control units, which effectively extend the dimensions of the barrier, or by curving the ends to increase the path length difference. An example of the prediction of

Example: Find the noise reduction due to a barrier of effective height 2 m, placed 2 m from a source when the receiver location is at 100 m from this.

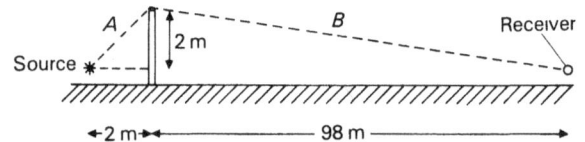

$$A = \sqrt{2^2 + 2^2} = 2\,8\ \text{m}$$

$$B = \sqrt{2^2 + 98^2} = 98\,0\ \text{m}$$

$$D = 100\ \text{m}$$

$$\delta = 100\,8 - 100 = 0\,8\ \text{m}$$

	Octave band center frequency (Hz)							
	63	125	250	500	1000	2000	4000	8000
PWL of machine source (Hz)	88	97	110	108	102	96	90	85

from $\text{SPL} = \text{PWL} - 20 \log r + \text{DI} - 11$

substituting $r = 100$ m; and for a hemisphere, DI $= +3$ dB

$$\text{SPL}_{100\,\text{m}} = \text{PWL} - 20 \log 100 + 3 - 11$$
$$= \text{PWL} - 48\ \text{dB}$$

SPL at 100 m without barrier (db)	40	49	62	60	54	48	42	37
Wavelength, λ (m)	5 4	2 7	1.4	0 69	0.34	0 17	0 086	0 043

from $N = 2\delta/\lambda$
substituting, $\delta = 0.8$ m

Fresnel number, N	0 3	0 6	1.1	2.3	4 7	9.3	18.6	37 2
Attenuation due to barrier (from figure 6 4) (dB)	10	12	13	17	19	23	24	24
SPL at 100 m with barrier (dB)	30	37	49	43	35	25	18	13

76

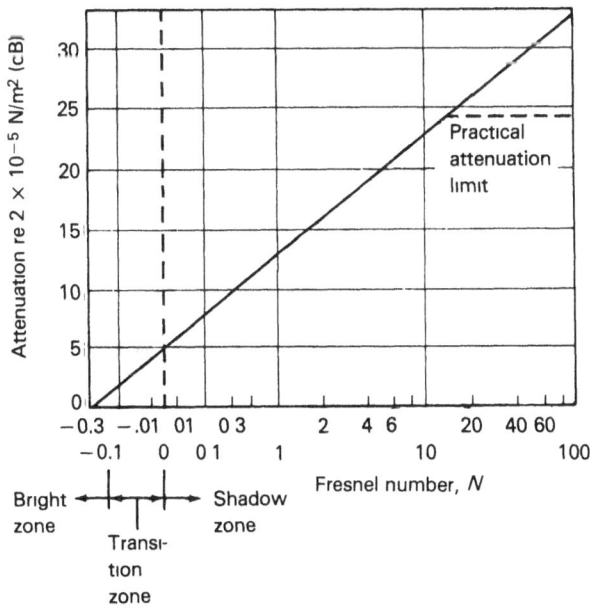

Figure 6.4 – Attenuation values for a barrier against Fresnel number (reference 7). Strictly for point sources. (From Leo L. Beranek, Noise and Vibration Control. Copyright © 1971 by McGraw-Hill, Inc. Used with the permission of McGraw-Hill Book Company.)

a barrier's performance calculated for each octave band is presented below. The attenuation values are taken from figure 6.4, which displays barrier attenuation as a function of Fresnel number, N.

Other nomograms are available[13] which enable the effect of β, the subtended angle of the barrier width at the source, to be taken into account. In addition, the angle of the barrier apex can be important, and double diffraction may even occur for particularly thick barriers.[14] One particularly detailed study involving considerable experimental work[15] considered the diffracted sound intensity around the top and around each edge and summed them to find the net barrier insertion loss, InL, according to:

$$\text{InL} = 10 \log (1/D)$$

and

$$D = \sum_i 1/(3 + 10N_i)$$

where N_i is the Fresnel number for the ith edge.

More complex expressions can be derived which include the separate effects of absorption on each side of the barrier, as well as the extent of the gap between the barrier and the ceiling.

6.2.1.2 Total enclosures

Complete enclosure is often the simplest and most effective procedure for solving a noise control problem when adequate source control is not possible. In particular, enclosures can serve a multipurpose role by providing a safety barrier, controlling lubricant and heat emission from a machine, and controlling the noise. It is usually easy to meet acoustic requirements this way, especially if the enclosure is vibration isolated from the machine and not too closely positioned to it. This minimizes the possibility of strong coupling effects caused by air resonances in the intervening space.

Typically, enclosure panels are specially designed for the situation; they often consist of a carefully balanced combination of absorbing, damping, and insulating materials incorporated into modular units which can be bolted to an angle-iron frame. The provision of absorption is important because it prevents high reverberant sound levels within the enclosure walls from degrading the overall insulating properties. In order to preclude flanking transmission caused by direct vibration of the enclosure walls, the machine source should be mounted on vibration isolators. Other factors affecting the acoustic performance include the potential weakening effect of apertures made to carry through essential services or for general access purposes. Where possible, service conduits should be grouped together and appropriate seals, such as lead-loaded vinyl, used to minimize the chance of serious leaks. Several examples of sealing techniques are shown in figures 6.5 and 6.6.

Double-glazed windows may be necessary where more than 20 dB(A) of attenuation is required; the general principles governing their performance are given in sections 5.2.2 and 5.2.3. Doors should be as massive as possible, with tight-fitting positive pressure seals around all edges. Adequate ventilation may be ensured by using specially integrated attenuating blower units or by having larger apertures with absorbent-lined baffles and bends as shown in figure 6.7. The cumulative effect of the heterogeneity introduced with all these elements is taken into account by computing the composite transmission loss, R, according to the method in section 5.2.3.

Enclosing a machine produces a reverberant field inside with an average sound pressure level, SPL_E, given by

$$\text{SPL}_E = \text{PWL} + 10 \log 4/R_{CE} \quad \text{dB}$$

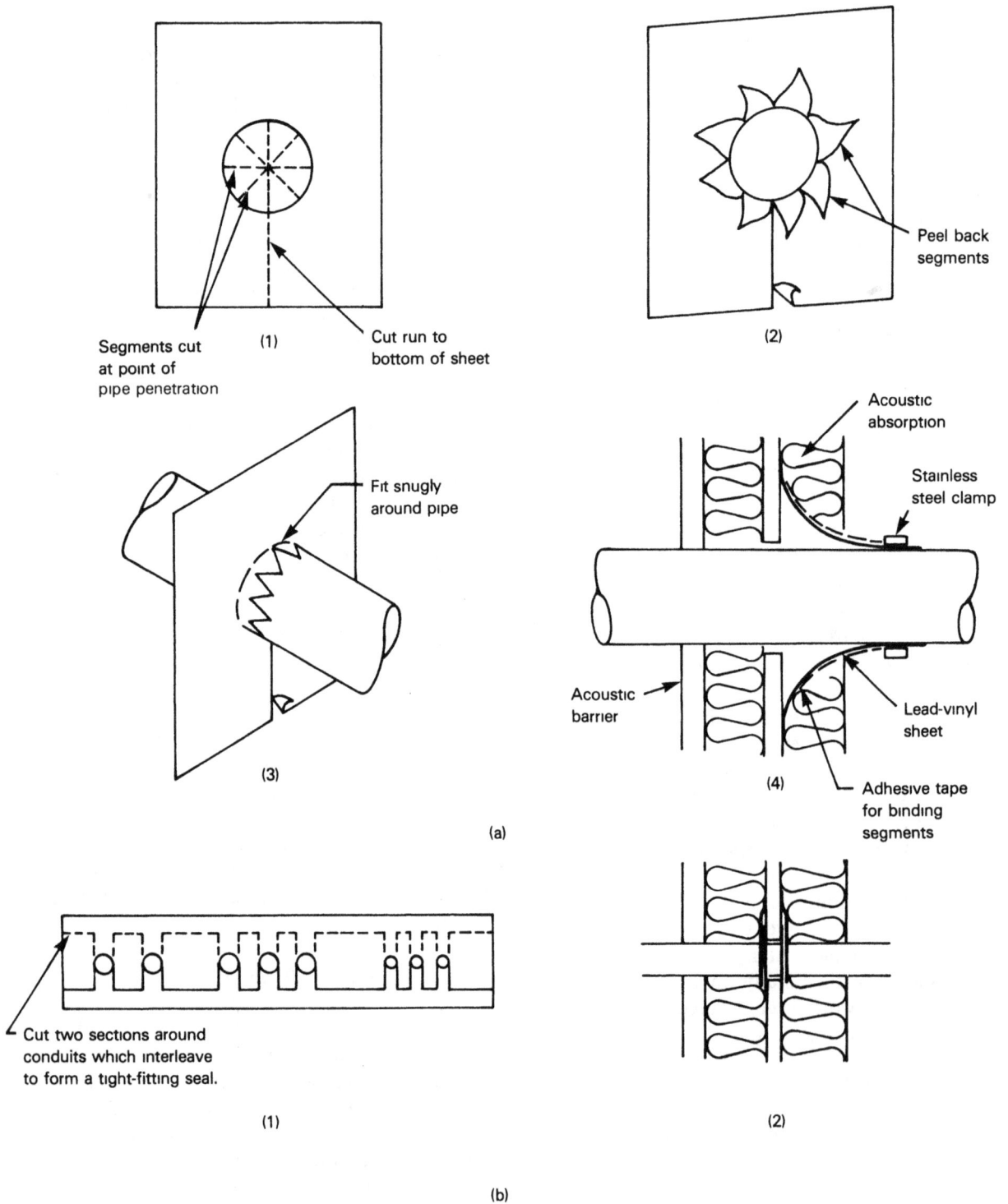

Figure 6.5 – Use of lead-vinyl sheet to seal acoustic leaks. (a) Changes of seal installation for a single large-bore pipe. (b) Installation of multi-seal for a series of small-bore pipes. (Derived from NASA Langley Research Center architectural drawings.)

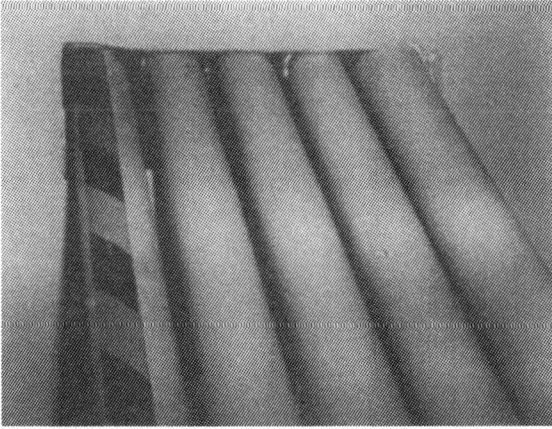

Figure 6.6 – Pipe multi-seal using lead vinyl sheet as insulation (by courtesy of NASA Langley Research Center).

where

PWL is the sound power level of machine source, dB

R_{CE} is the room constant of enclosure interior ($= S_E \overline{\alpha}_E / 1 - \overline{\alpha}_E$), m^2.

The level just outside the enclosure wall, SPL_{EXT}, is given by

$$SPL_{EXT} = SPL_E - R_E - 6 \quad dB$$

where R_E is the average transmission loss of enclosure walls, dB.

The net effect is to replace the machine source with an enclosure of total sound power level PWL_E given by

$$PWL_E = SPL_{EXT} + 10 \log S_E \quad dB$$

where S_E is the radiating surface of enclosure, m^2.

From this the general reverberant level SPL_{REV} in a room resulting from an enclosed machine noise source is given by

$$SPL_{REV} = PWL_E + 10 \log 4/R_C \quad dB$$

where R_C is the room constant of the room, m^2.

These expressions may be summarized in the relationship below, which is analogous to that presented in section 6.2.1:

$$SPL_{REV} = SPL_E - R_E + 10 \log S_E \\ - 10 \log A \quad dB$$

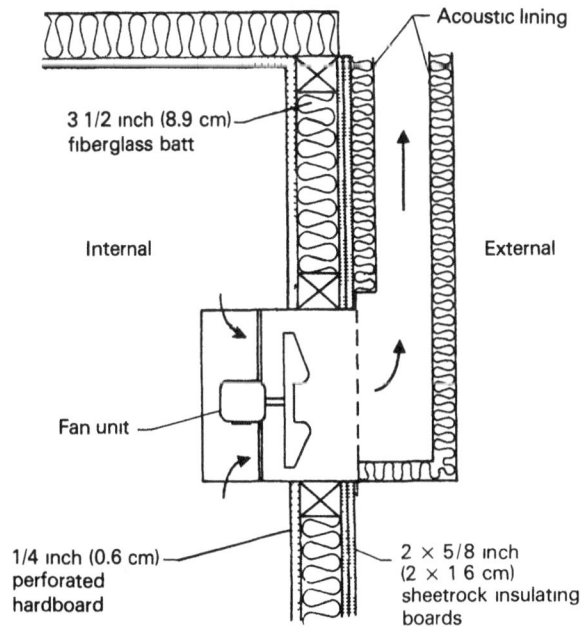

Figure 6.7 – Example of an assisted ventilation enclosure aperture. (Derived from NASA Langley Research Center architectural drawings.)

where A is the total absorption within the room ($S\overline{\alpha}$), assumed approximately equal to R_C.

Note: This relationship assumes a reverberant forcing field inside the enclosure, omnidirectional source radiation, and a common value of R for all surfaces.

Figure 6.8 shows before and after photographs of an enclosure around an air handling unit, highlighting the complexities of placing an enclosure around such a unit with a large array of pipe and service connections. An example of calculating the performance of an enclosure follows:

Example: A generator with known sound power levels is enclosed within a structure whose dimensions are $3 \times 3 \times 2.5$ m. The average transmission loss of the enclosure walls, R, the mean absorption coefficient within the enclosure, $\overline{\alpha}_E$, and total room absorption, A, are given in octave bands.

It is assumed that the generator inside the enclosure does not affect $\overline{\alpha}_E$, either by obscuring or adding to the existing absorption, and that all enclosure surfaces radiate omnidirectionally.

The results are presented as a comparison of A-weighted reverberant levels within the room, with and without the enclosure.

79

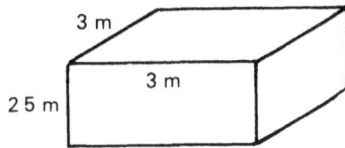

Total internal surface area, $S = 48$ m^2
Total radiating surface area, $S_E = 39$ m^2

The example highlights the importance of using sufficient absorption inside the enclosure, especially when the low frequency power level is relatively high. In the calculation it may be noted that the enclosure resulted in very little attenuation at 63 Hz.

	Octave band center frequency (Hz)						
	63	125	250	500	1000	2000	4000
Enclosure mean absorption coefficient, $\bar{\alpha}_E$	0 14	0 20	0 30	0 63	0 91	0 91	0 88

From $R_{CE} = S\bar{\alpha}_E/1 - \bar{\alpha}_E$
and substituting $S = 48$ m^2, and $\bar{\alpha}_E$

Enclosure room constant, R_{CE} (m^2)	7 8	12 0	20 6	81 7	485 3	485 3	352 0
10 log $4/R_{CE}$ (dB)	−2 9	−4 8	−7 1	−13 1	−20 8	−20 8	−19 4
PWL of generator (dB)	109	109	109	103	100	92	84

From $SPL_E = PWL + 10 \log 4/R_{CE}$

SPL$_E$ inside enclosure (dB)	106 1	104 2	101 9	89 9	79 2	71 2	64 6
Average transmission loss of enclosure walls, $\bar{R}_E$ (dB)	8	13	18	23	28	33	35
Total absorption of room, A (dB)	30	30	30	60	60	60	90

From $SPL_{REV} = SPL_E - \bar{R}_E + 10 \log S_E$
$- 10 \log A$
and substituting $S_E = 39$ m^2, and for SPL$_E$, $\bar{R}_E$,
and A

SPL$_{REV}$ in room with enclosure (dB)	99 2	92 3	85 0	65 0	49 3	36 3	26 0

From $SPL_{REV} = PWL + 10 \log 4/A$

SPL$_{REV}$ in room without enclosure (dB)	100 2	100 2	100 2	91 2	88 2	80 2	70 5
A-weighting	−26	−16	−8 5	−3	0	+1	−1
SPL$_{REV}$ with enclosure (dB(A))	73 2	76 3	76 5	62	49 3	37 3	+25
SPL$_{REV}$ without enclosure (dB(A))	74 2	84 2	92 7	88 2	88 2	81 2	69 5

Hence, overall sound pressure level in room with enclosure = 80 dB(A)
overall sound pressure level in room without enclosure = 95 dB(A)

(a)

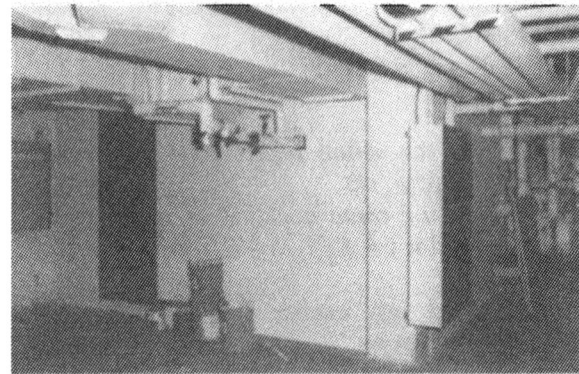

(b)

Figure 6.8 – Example of enclosure fitted around an air handling unit. (a) Unit without enclosure. (b) Similar unit after enclosure. (By courtesy of NASA Langley Research Center.)

The predicted insertion loss values of the enclosure walls, i.e., the difference between the room overall sound pressure levels with and without the enclosure, must represent something close to the actual values obtained in a practical installation. For instance, it has been shown that some foam absorbers can have a pronounced dilatational resonance,[16] which can seriously detract from the performance of the overall partition. This particular problem can be solved by having the foam loosely attached to the insulating panel in only a few places. For particularly critical installations, it may be of some advantage to use a computer program developed under NASA sponsorship[17] which predicts panel response to reverberant acoustic fields. The required program input parameters are the power spectral density of the sound field, panel modes of concern and the associated damping values, as well as the definition of the panel boundary conditions. The program then furnishes the fundamental resonance frequencies and associated mode shapes; it may be easily adapted to handle other types of acoustic fields as well.

When enclosures are built very close to the source of noise, such that the separation distance is less than half the wavelength of the lowest frequency of interest, then the effect of the intervening air space becomes important. Resonances in the cavity space may exhibit a strong coupling effect with panel resonances; furthermore, the assumption of a reverberant field inside the enclosure is questionable when the machine occupies a large fraction of the enclosure volume. This leads to substantial deviations from simple mass law predictions, which are not easily accounted for.[18,19]

Recently the problem of predicting the acoustic environment within the space shuttle payload bay has resulted in the development of an analytical model[20] to predict space-averaged levels within the empty or partly filled bay. This incorporates the use of statistical energy analysis, SEA, to be used when the excitation field is diffuse and there are large numbers of acoustic and structural modes present.[21] At low frequencies, account must be taken of each individual mode, whereas SEA utilizes the concept of an average modal density in frequencies (see section 6.2.4). The model computes net power flow from the exterior to the interior and between subvolumes surrounding the payload. The potential value of this model with respect to other applications is in accounting for variations in sound pressure level produced when large-scale structures are introduced into a cavity: these are particularly evident at low frequencies. Some of these theoretical results have been validated using physical scale models.[20,22] The most difficult task in this approach is to define adequately a series of subvolumes, each of which must be considered homogeneous in terms of the acoustic field. A finite element approach can also be used.[23]

A comprehensive review[24] of all the recent accomplishments in the prediction of interior noise for transportation vehicles and cavities of various sorts helps to summarize the work of NASA in this field. It is considered therein that the increasing use of modal representations of walls and cavities provides a useful conceptual and computational framework with which to improve existing structural noise transmission prediction

methods. The convenience of avoiding a determination of each separate natural mode when using SEA is pointed out, as well as the need for less detailed modelling of acoustic modes when there is appreciable absorption present.

6.2.1.3 Partial enclosures

Partial enclosures may simply be treated as enclosures with holes or missing panels. The interior absorption coefficient α and transmission coefficient τ should both be set equal to unity where there are open sides. Because the assumption of a reverberant field may often still be made, normal enclosure prediction formulas as given in section 6.2.1.2 remain valid with the above coefficient substitutions. Partial enclosure performance, however, is obviously limited and can be difficult to predict because of serious directivity effects at the open sides. If major noise sources are kept well within the enclosure and their spectrum is mainly high frequency, then a reduction of as much as 10 dB in level may be gained outside the open end. This is more readily achieved by the use of internal absorption, especially with judicious positioning of this near the open areas. Figure 6.9 presents some partial barrier configurations and their anticipated performance.

Partial enclosures can be particularly useful where there are severe access constraints to a machine because of supply/product feed or ventilation requirements. Partial enclosures may also be helpful in reducing localized noise sources which are integral parts of a larger machine surface.

6.2.2 Use of Acoustic Absorption in Reverberation Control

Absorption has limited value within the direct field of a source unless used in conjunction with insulating materials in enclosures. Figure 5.9 shows the appropriate placement of material for different propagation paths. As a general rule, at less than 3 m from machine surfaces, no significant reduction may be achieved with absorption alone. The transition from a direct to a reverberant field occurs at an approximate distance r_c (see figure 3.11), which may be computed according to

$$r_c \approx 0.2 \sqrt{S\bar{\alpha}} \;\; \text{m}$$

It is sometimes possible to detect this changeover aurally by cupping the hands behind

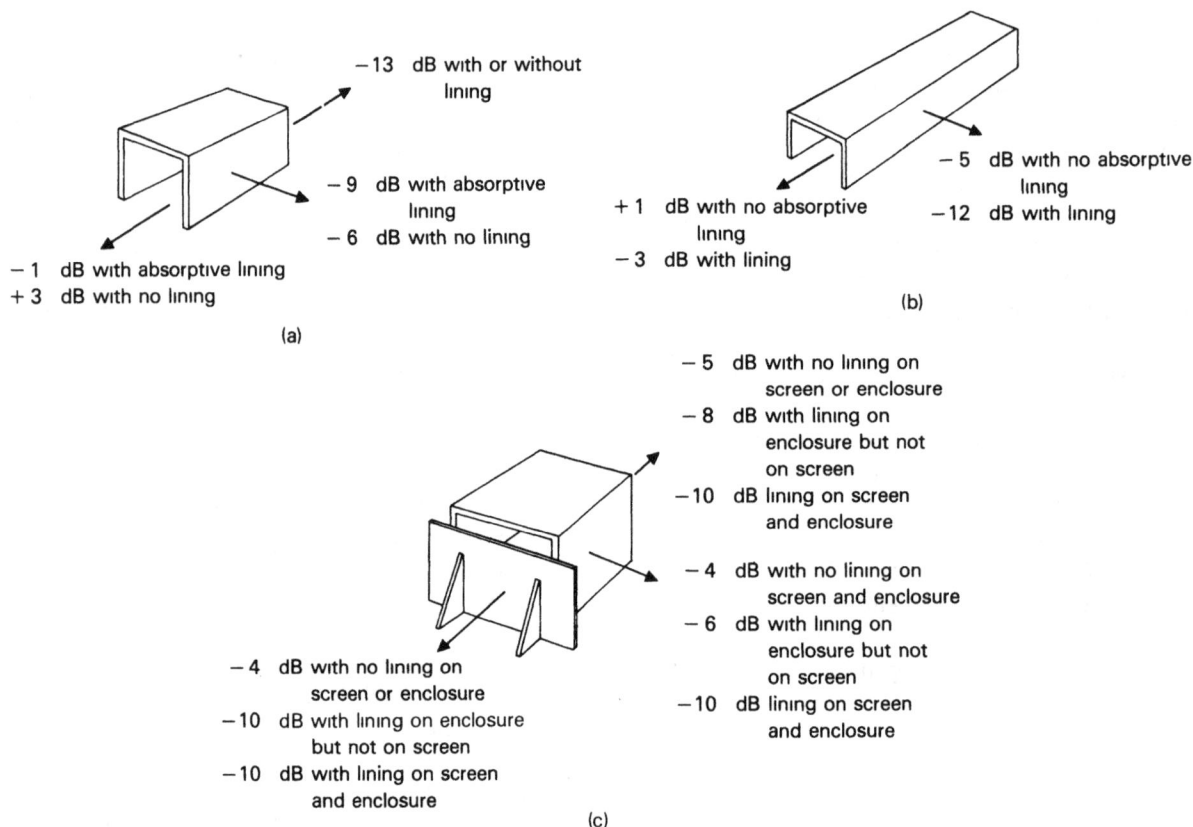

Figure 6.9 - Partial enclosure approximate attenuation performances. Values specify average change in SPL in the frequency range 500–4000 Hz. (a) Enclosure with one side open. (b) Tunnel. (c) Open-sided enclosure and moveable screen. (From Ian Sharland, Woods Practical Guide to Noise Control, *1972. Used with the permission of Woods Acoustics.)*

the ears and finding the position for which any detectable source directionality is seriously diminished. Normally the noise exposure of a machine's operator is dominated by the direct sound field. However, in some instances, noisy machines may produce reverberant sound levels which exceed the direct field levels received by operators of quieter machines. Some examples of this situation and the effects of possible treatments illustrating the balance between reverberant and direct fields are shown in figure 6.10. When sources or machines are scattered over the whole floor area, it is unusual for very few positions to be outside the dominant influence of direct fields.

The usual objectives of reverberant field control are to reduce the noise exposure of personnel below the statutory or desired limits, to improve speech communications and the detection of aural warning signals, and to reduce net incident sound pressure level, thereby alleviating the required per-

formance of insulating materials. It is commonly accepted that the improvement from small reductions in the reverberant level are subjectively rather greater than would be anticipated merely from consideration of the changes in the overall sound pressure level. It should be cautioned, however, that only rarely are sound fields truly diffuse; consequently there is often considerable uncertainty as to the exact reduction which is achievable at a particular point by means of extra absorption. This is especially so where the room shape deviates from regular geometry as, for example, in very long, low-ceilinged buildings.

As noted in section 6.2.1, the reverberation level is related to room characteristics by

$$SPL = PWL + 10 \log 4/R_c \quad dB$$

where R_c is the room constant ($= S\bar{\alpha}/1 - \bar{\alpha}$), m^2.

The quantity $S\bar{\alpha}$, which is a measure of the total absorption in the room, is found by assessing

Before

Action
Required

After

+10

3 dB

+10

Reduction in total
SPL of 3 dB

0

0

0 dB

Reduction in
reverberant SPL
by 20 dB

−10

−10

−20

−20

+10

+10

0

0 dB

Reduction in
direct SPL
by 20 dB

Reduction in total
SPL of 10 dB

−10

−10

−20

−20

+10

+10

0

0 dB

Reduction in
reverberant SPL
of 10 dB

Reduction in total
SPL of 7 dB

−10

−10

−20

−20

Relative SPL re 2 × 10⁻⁵ N/m² (dB)

Total SPL

Direct-field SPL

Reverberant-field SPL

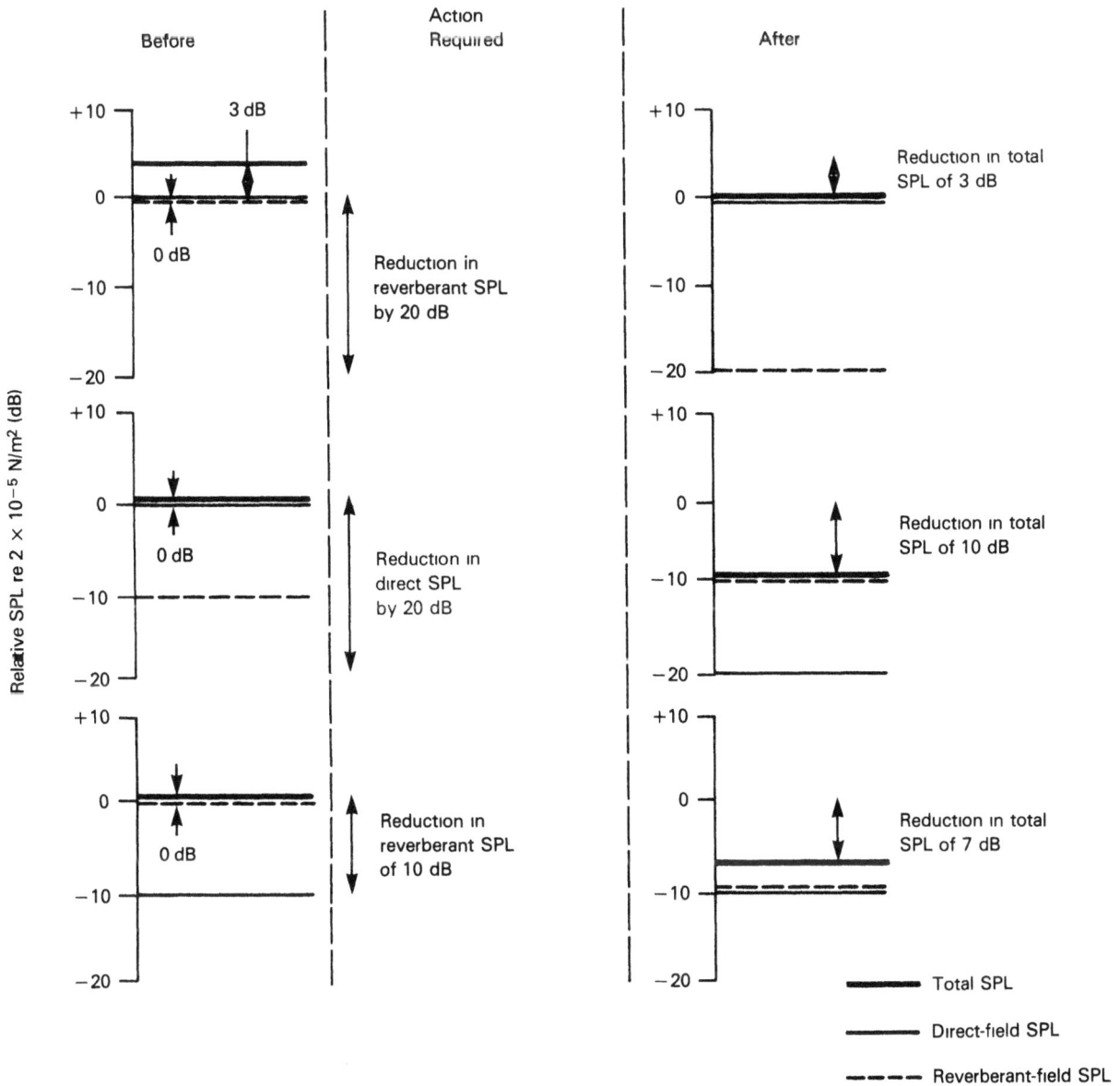

Figure 6.10 – Contributions of direct and reverberant fields to the total sound pressure level.

the reverberation time as defined by Sabine (see below). This is the time taken for a steady-state reverberant sound field to decay to 60 dB below its initial level after removing the noise source input. This is usually measured using a starting pistol or a high level broadband source, and the decay rate is measured in each octave band. The following equation may be used except where the room is extremely large, in which case air absorption of high frequencies becomes important:

$$\text{Reverberation time, } T_R = \frac{0.16V}{S\overline{\alpha}} \text{ s}$$

where

V is the room volume, m³
$S\overline{\alpha}$ is the total absorption for i areas with separate absorption characteristics ($= \Sigma S_i \alpha_i$), metric sabins, m².

Note: A separate term dependent on room volume and frequency is added to the denominator to account for air absorption when necessary.

There may be several difficulties even in making this simple estimate, because of uncertainties in estimating the total volume V and the internal surface area S of many industrial spaces.[25] Also, high ambient noise levels tend to make suitable measurements of the decay curve difficult and the inherent assumption of a diffuse field is often invalid. Figure 6.11a shows a typical change in reverberation time for varying amounts of absorption added to the main doors, ceilings, and upper half of the walls of a large machine shop. In this case, the precaution was taken of averaging the results at three separate locations to account for spatial variations. The accompanying inset table shows the actual performance against the design goal. A photograph of the absorption treatment, which was spray-installed using a cellulose base of average depth 3.0 in. (7.6 cm), is shown in figure 6.11b.

The reduction in sound level ΔSPL which can be expected from the addition of absorption material is related to the room constant by

$$\Delta SPL = 10 \log R_{c2}/R_{c1} \quad \text{dB}$$

where

R_{c1} is the room constant before absorption treatment

R_{c2} is the room constant after absorption treatment.

A ready account of the attenuation may be made with reference to figure 6.12, from which it may be seen that there is a decreasing benefit in increasing the absorption without limit. As a rough guide, a factor of 5 change in the average absorption coefficient $\overline{\alpha}$ produces a change of 10 dB or so in the sound pressure level.

It is often possible to add appreciable absorption only in the ceiling area. This may conveniently be achieved by the use of functional absorbers (see section 5.1.3 and figures 5.9 and 5.10). Other useful techniques include standard modular absorbing panels, using a spray-on fibrous absorbent which can often be applied

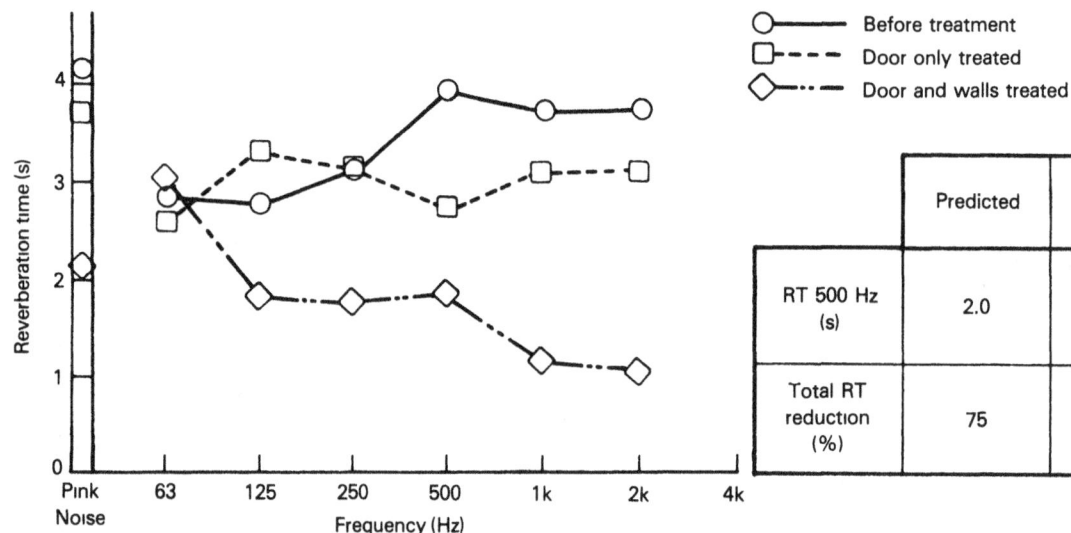

	Predicted	Achieved
RT 500 Hz (s)	2.0	1.9
Total RT reduction (%)	75	49

a)

(b)

Figure 6.11 – Reverberation field control in a large industrial space. (a) Reverberation time changes. (b) Photographs of installed absorption. (By courtesy of NASA Langley Research Center.)

easily and economically, and resonator units for handling discrete frequency problems (see figure 5.6).

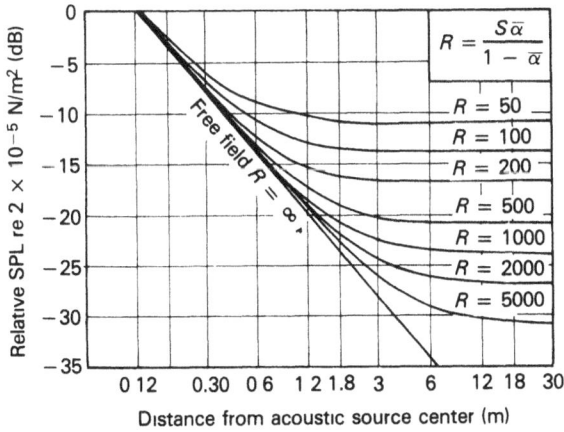

Figure 6.12 – Attenuation in reverberation field with change in room constant and the transition effect from a direct field regime.

6.2.3 Flanking Path Analysis and Identification

Energy not transmitted from the source to the receiver along the most direct path is termed flanking or indirect energy. This may occur as a result of mechanical excitation of the structure and subsequent re-radiation at points distant from the original source; break-out of energy from ducts or ceiling voids into neighboring spaces; and airborne sound transmission through weak points in existing insulating barriers. Several of these routes are illustrated in figure 6.13.

It is not easy to determine the extent of flanking transmission in many situations. As a general rule, flanking becomes significant where insulation above 35 dB is desired, and becomes particularly critical where 50 dB of insulation and over is required.[26] To achieve insulation above this, it is usually necessary to build a completely isolated inner room. There are several techniques which can be used to estimate the extent of the problem. A substantial difference between the predicted and observed values for the transmission

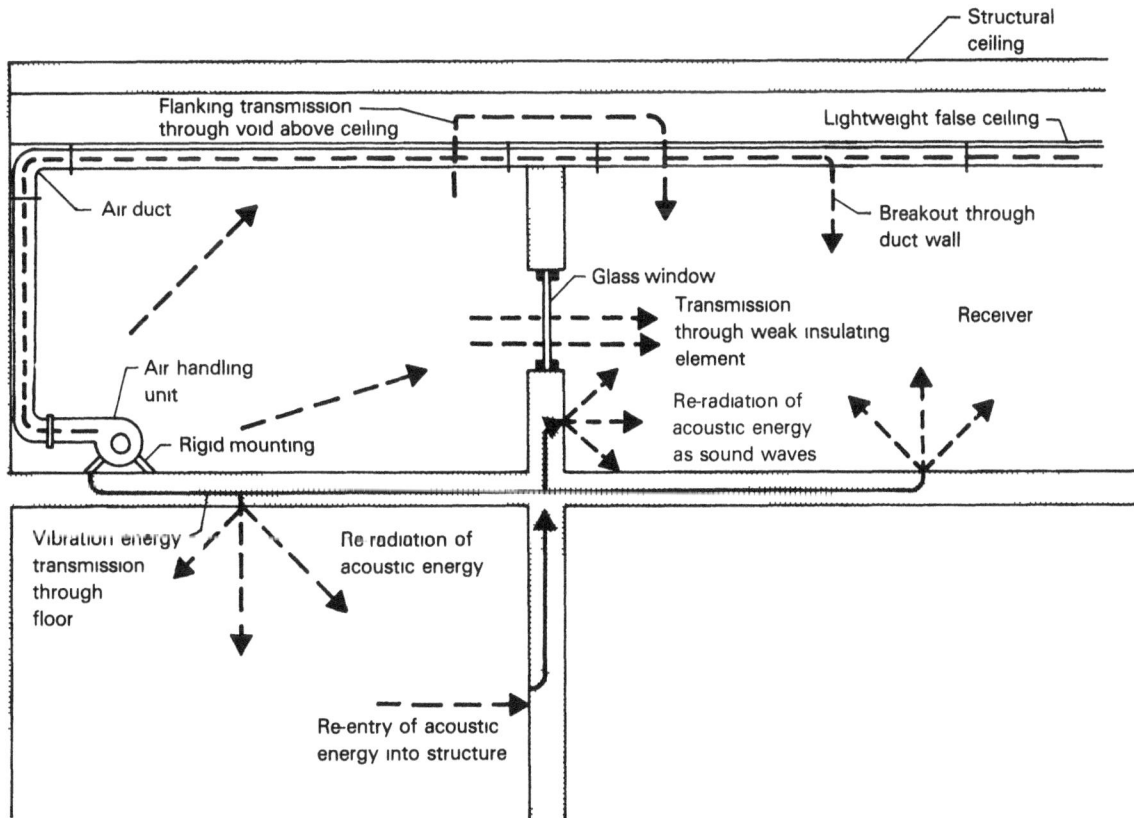

Figure 6.13 – Flanking and indirect transmission paths.

loss R, of the order of 5 dB or more, often indicates the amount of sound propagation via unaccounted paths. In this case the slope of the octave band plot for the partition insulation is often less than predicted. Directional microphones are sometimes useful for locating areas radiating into a room when the acoustic space concerned is relatively small and anechoic. Even more effective is the direct assessment of sound power radiated from neighboring walls relative to the insulating partition of interest, using accelerometers. The appropriate equation for estimating this is

$$W = \sigma <\bar{v}^2> \varrho c S \quad W$$

where

σ is the radiation efficiency of the panel, usually approximately 1

$<\bar{v}^2>$ is the space averaged mean square velocity of panel surface, m/s

ϱc is the specific acoustic impedance of air, N·s/m

S is the radiating area of panel, m^2.

Note: This equation is strictly applicable only above the panel critical frequency f_c.

Other techniques employed for this purpose include correlation analysis (see section 3.4.1), and the use of a vibration isolated noise source (e.g., a loudspeaker) in place of a nonisolated source. The latter procedure enables a comparison to be made between the attenuation obtained with the substitute source and that obtained with the actual source. Any disparity can be taken as an estimate of structure-borne transmission.

One particular approach used to account comprehensively for flanking paths stemmed from the development of a series of equations for coping with multiple path propagation in a helicopter internal noise study.[27] This may be suitable for well-defined noise control problems. The general procedure is applicable to other internal noise prediction cases where structural geometries, noise source strengths, and material acoustic properties are defined. Because of the importance of minimizing the weight penalty associated with acoustic treatment in this particular situation, the trade-off between insulation and absorption was emphasized. For the case of sound propagation through $N-1$ cavities sharing common boundaries with cavity N, the sound pressure level in the latter is given by

$$SPL_N = PWL - 10 \log S_s + 10 \log$$

$$\left[(C_0 + 4/R_0)(C_1 + 4/R_1) \cdots (C_N + 4/R_N) \right]$$

$$+ 10 \log \left[(\tau_0)(\tau_1)(\tau_2) \cdots (\tau_{N-1}) \right] \quad dB$$

where

PWL is the source power level, dB

S_s is the source area, m^2

R_N is the room constant of the Nth cavity, m^2

C_N is the correction factor of the Nth cavity, dependent on source shape and distance and orientation of source with respect to receiver.

τ_{N-1} is the transmission coefficient associated with the partition between the $N-1$th and Nth cavities.

Comparing the relative magnitude of C_N with $4/R_N$ indicates whether the direct or reverberant field dominates. As the cavity walls become less effective as barriers, then the product of τ's tends to unity, and consequently zero net barrier attenuation results, since $\log (1) = 0$.

For the case of structural radiation by N surfaces into a single cavity, the resultant sound pressure level is given by

$$SPL = PWL_1 - 10 \log S_1 + 10 \log$$

$$\left[\tau_1(C_1 + 4/R) + \frac{\tau_2(C_2 + 4/R)}{\beta_2/\epsilon_2} \right.$$

$$\left. + \cdots + \frac{\tau_N(C_N + 4/R)}{\beta_N/\epsilon_N} \right] \quad dB$$

where

PWL_1 is the sound power level at wall 1

S_1 is the area of wall 1

β_N is the area ratio S_N/S_1

ϵ_N is the power ratio W_N/W_1.

Note: ϵ_N/β_N is the ratio of the intensity at wall N to the intensity at wall 1.

The study[27] elucidates an integrated step-by-step procedure utilizing the above equations to obtain the total sound pressure level at a particular space after accounting for structural and geometrical factors, major source strengths and paths, transmission loss and absorption data, correction factors from radiating surface sizes, and distances to the receiver position.

Some of the procedures to reduce weakened transmission paths have already been dealt with in the section on enclosures (section 6.2.1.2). Good caulking at panel edge and apertures with bitumastic-type compounds, positive pressure seals around doors, double panels resiliently isolated, resilient mounting of ducts and service connections, and vibration isolation of machines and enclosures can all help achieve the requisite insulation performance without degradation from unaccounted propagation routes.

For the helicopter problem a wide variety of remedial procedures were used to reduce the interior level by some 29 dB(A).[28] These included applying damping tape to structural members, mounting the interior trim on antivibration isolators to preclude strong radiation from the sides and ceiling, acoustically sealing the doors and ducts, installing a raised plywood floor 2 cm thick, and filling the volume between trim and fuselage walls with 16-kg/m^3 density fiberglass or two lead layers separated with absorbent foam of total thickness 5 cm and total surface density 7.3 kg/m^2. These are summarized in figure 6.14, which shows the acoustic treatment configurations as well as major noise sources.

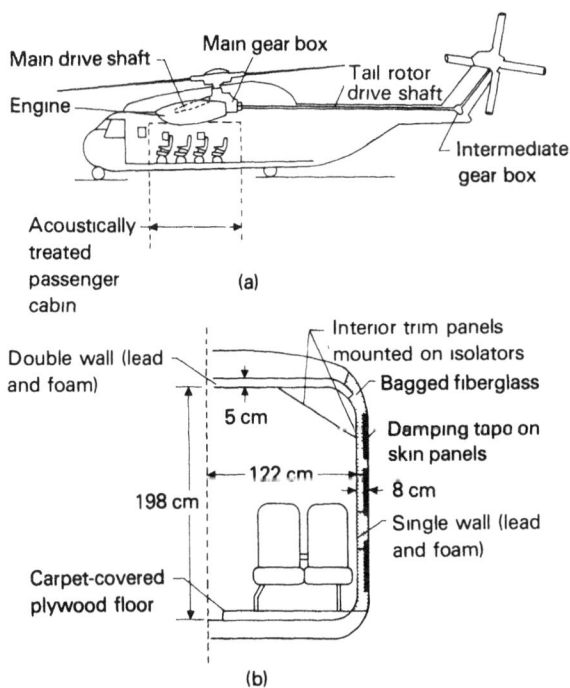

Figure 6.14 – Illustrative sketches of integrated acoustic treatment in a helicopter noise problem (reference 28). (a) Side view. (b) Internal cross section.

6.2.4 Statistical Energy Analysis (SEA)

In the early 1960s statistical energy analysis evolved mainly as a result of NASA and USAF funding of studies regarding complex noise and vibration problems in high speed aircraft and spacecraft. The three main areas of application have been structural vibration excited by airborne noise, generation of a sound field within an elastic container caused by external excitation by noise, and the flux of vibratory energy through interconnected elastic structures.[29] Many of the original principles were developed from room acoustics modal analysis.

Reference 30 summarizes the early work on acoustic excitation of structures and provides a useful introduction to the techniques involved. Basically, SEA relies on estimating the average number of resonances in a moderately broad band of frequencies and the spatial distribution of response amplitude for a "typical" mode of resonance. By computing the average response of this typical mode to sound waves from many angles, it is possible to estimate the average space-time response in that frequency band, after multiplication with the appropriate modal density. The benefit of such an approach is that, even if all the individual modes could be exactly characterized, the computation of the separate effects of each mode would be impossibly cumbersome.

Normally at least two of three distinct phases are involved in conducting SEA analysis, namely,[31] the development of a power-flow model with component areas of homogeneous energy distribution; evaluation of parameters such as modal densities, structural loss factors, and coupling loss factors; and the calculation of equilibrium energy distribution and the corresponding statistical values of variables such as plate velocity and acoustic pressure, which are of direct concern to the engineer. It is important to recognize that SEA methods are inherently inappropriate at low frequencies, where statistical averaging cannot be applied because modal density is insufficient and where the coherent addition of modes (see section 3.2) and mode-to-mode coupling can become important.

A typical representation of an SEA model applied to sound transmission across a single panel is shown in figure 6.15. The loss factor η_i and the coupling loss factor η_{ij}, both dimensionless quantities, are defined as[32, 33]

$$\eta_i = \frac{\Pi_{i,\text{diss}}}{\omega E_{\text{stored}}}$$

where

$\Pi_{i,\text{diss}}$ is the power dissipated in the ith element, W

ω is the angular frequency $(= 2\pi f)$, Hz

E_{stored} is the stored energy, J;

$$\eta_{ij} = \frac{\Pi_{ij}}{\omega(E_i - E_j)}$$

where

Π_{ij} is the power flow between systems i and j, W

E_i, E_j is the total energy of systems i and j, respectively, J.

The total modal count in the ith system, N_i, is related to the modal density n_i by

$$N_i = n_i \Delta f$$

where Δf is the frequency bandwidth of interest.

A series of simultaneous equations is then written and subsequently solved:

$$\Pi_{1,\text{in}} = \Pi_{1,\text{diss}} + \Pi_{12} + \Pi_{13}$$

$$\Pi_{2,\text{in}} = \Pi_{2,\text{diss}} - \Pi_{12} + \Pi_{23}$$

$$\Pi_{3,\text{in}} = \Pi_{3,\text{diss}} - \Pi_{13} - \Pi_{23}$$

or alternatively,

$$\Pi_{1,\text{in}} = \omega\eta_1 E_1 + \omega\eta_{12}\, n_1 \left(\frac{E_1}{n_1} - \frac{E_2}{n_2}\right) +$$

$$\omega\eta_{13} n_1 \left(\frac{E_1}{n_1} - \frac{E_3}{n_3}\right) \quad \text{etc.}$$

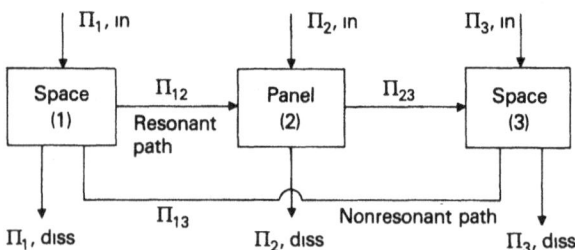

Figure 6.15 – Single-panel sea model.

Although SEA has been used mainly as a research tool, these techniques are used increasingly in building acoustics, marine engineering and shipbuilding, and nuclear reactor design, as well as in industrial noise problems. An example of the predictive accuracy with SEA appears in figure 6.16, which is taken from a study on acoustic excitation of a spacecraft shroud and the subsequent effect on the interior space.[34] The calculated noise reduction is consistently lower at low frequencies, probably as a result of the reduction in internal acoustic modal density.

Fortunately, some relevant parameters, such as acoustical radiation impedance, which provides a measure of the coupling between sound fields, and modal densities, can be calculated using basic acoustical theory.[35]

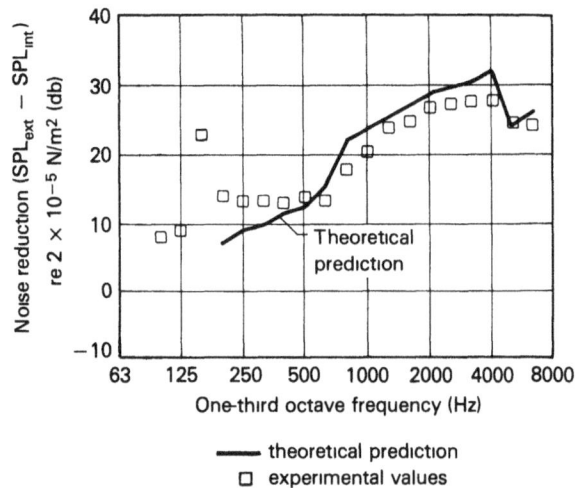

Figure 6.16 – Comparison of experimental and theoretical (SEA) noise reduction for a noise transmission problem (reference 34).

6.3 ADMINISTRATIVE AND HEARING PROTECTION PROGRAMS FOR RECEIVERS

After suitable assessment of noise exposure parameters, the risk of hearing hazard should be evaluated. In critical work positions, it is usually best to use noise dosimeters to check on the total exposure received by each employee. To comply with OSHA regulations, the allowed exposure dose for each 8-hour period may not exceed unity, which is equivalent to a 90 dB(A) continuous level (see the Appendix). If possible, the exposure dose should not exceed 0.5, to obviate the need for any form of hearing protection. As shown in figure A.1 in the Appendix, levels over 85 dB(A) carry a serious risk of eventual permanent hearing damage. It is important therefore to develop and implement source and path engineering solutions to noise control. Control at the receiver position should not be considered unless source and path engineering solutions require a substantial lead time for complete installation or unless they can provide only a limited attenuation. The following sections set forth the long-term limitations of administrative or personal hearing protection programs, from the standpoint of economics and of satisfactory acoustic attenuation.

6.3.1 Personal Hearing Protection

There are two common types of protectors, namely, ear plugs and ear muffs, with several in-termediate varieties as well. The former are often prefabricated in several sizes, although some may be custom-shaped in situ for the wearer. Ear plugs made of glass down or waxed cotton wool, designed to be used only once, are also available. The attenuation afforded by these types varies widely, depending on the manufacturer, wearer, and quality of fit. Consequently, both the average attenuation and the standard deviation over a range of people are usually quoted to provide a better estimate of their potential value. Earmuffs consist of circumaural cups, usually built of plastic shells with fluid- or foam-filled seals, fitting completely around each ear and connected by a properly tensioned headband. Photographs of the various types of hearing protectors are shown in figure 6.17.

The performance of ear plugs is more variable than that of muffs because plugs require an individual optimization for each wearer; moreover, workers often insert plugs incompletely because

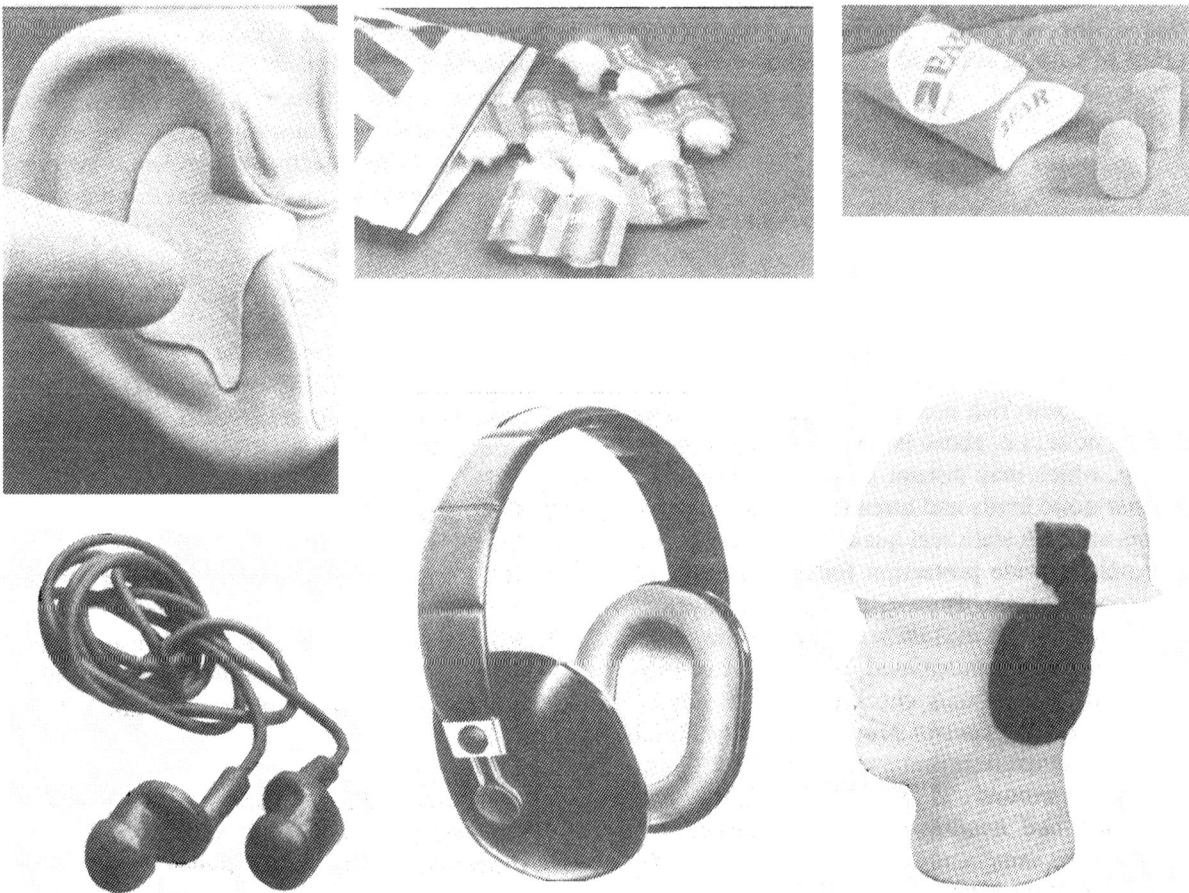

Figure 6.17 – Examples of common types of hearing protectors. (By courtesy of Flents Products Co., Bilsom International, Inc., Adco Hearing Conservation, Inc., and E.A.R. Corp.)

they may be uncomfortable. The performance of ear muffs depends on correct tension in the head-band, proper maintenance of the seals around the cups, and the absence of long hair or spectacles, which will detract from their correct fitting. Two standard procedures for testing the performance of either type are now available,[36, 37] and the Environmental Protection Agency has proposed a single number rating, based in part on the testing procedure set forth in reference 37, called the noise reduction rating (NRR).[38] The NRR is sub-tracted from a dB(C) estimate of the noise climate to give the net level received at the ear in dB(A). It is intended to introduce this procedure for rating and labelling all future hearing protectors. The National Institute for Occupational Safety and Health (NIOSH) has published a comprehensive list of attenuation data and standard deviations for proprietary hearing protector models.[39] A recent report[40] has also established that the noise attenuation provided by ear plugs in actual use is much less than that predicted by laboratory test-ing. This appears to be the result of improper fit-ting in the workplace, underscoring the impor-tance of careful selection and skilled supervision when ear plugs are initially introduced to employees.

6.3.2 Monitoring Audiometry and Administrative Control Procedures

In conjunction with the use of hearing protec-tors, a monitoring audiometry program should be implemented, with semiannual hearing tests for employees who risk hearing damage from a high level of noise, i.e. above 90 dB(A). This expensive testing, which may disrupt production, is needed because noise limits and attenuation data for pro-tectors are both statistical quantities and can there-fore only provide protection for a percentage of an exposed group. In addition, since protectors are often removed, and since noise levels may fluc-tuate more than anticipated, a monitoring pro-gram provides the only check on the efficacy of the overall noise control program. A substantial survey of companies conducted by NIOSH[41] indi-cates that around 25 percent of those who responded had hearing conservation programs; however, the same study also showed severe defi-ciencies in audiometric testing equipment and in the training and qualifications of personnel who operate them.

A hearing conservation strategy should also include a continual education program with such features as labelling of hazardous areas with warn-ing signs, personal instruction for fitting hearing protectors, informative films and pamphlets, and personal example by management and union offi-cials. All new employees should be given instruc-tion on hearing conservation as part of their formal induction schedule.

Administrative noise control procedures are based on the principle that exposure is a function both of time and level. According to the OSHA standard (see the Appendix), the trade-off be-tween these is a 5 dB increase in permissible level for each halving of the exposure time. For exam-ple, employees who steadily experience 90 dB(A) should never be exposed to higher levels, while those who do have exposure above 90 dB(A) should be removed from the noise source after expiration of the relevant OSHA time limit and should spend the balance of the day in areas at under 90 dB(A). This approach obviously requires careful allocation and scheduling of work duties; in practice it has met with little success.[41] It works best where it is feasible to split shifts, so that an individual's work time is divided between operations at various noise levels, or by using a particularly noisy machine on several different days rather than on one occasional full day.

REFERENCES

1. Eckert, W. L., E. T. Booth, P. D. Emerson and J. R. Bailey, Fly-Shuttle Loom Noise Source Identification, ASME 76-DE-39, April 1976.

2. Bergmann, E. P., Severe Sources of Industrial Machinery Noise (EPA Contract 68-01-2234, IIT Research Institute), Task Report J 6331, November 1974.

3. Royster, L. H., and J. R. Stephenson, Char-acteristics of Several Industrial Noise Envir-onments, J. Sound Vib., vol. 47, no. 3, 1976, p. 313.

4. L. S. Goodfriend Associates, Noise from Industrial Plants (prepared for EPA), availa-ble from NTIS as NTID300.2, 1971.

5. Emerson, P. D., J. R. Bailey, and W. D. Cooper, Economic Impact of a 90 dB(A) Noise Standard on Textile Spinning Opera-tions, ASME 77-RC-15, May 1977.

6. Miller, T. D., Industrial Noise Control: Putting it all Together, Noise Con. Eng., vol. 9, no. 1, July–August 1977, p. 24.

7. Beranek, Leo L., ed., *Noise and Vibration Control,* McGraw-Hill Book Co., 1971.

8. Stewart, J. S., and F. D. Hart, Control of Industrial Wood Planer Noise through Improved Cutterhead Design, Noise Con. Eng., vol. 7, no. 1, July–August 1976, p. 4.

9. Stewart, J. S., A Theoretical and Experimental Study of Wood Planer Noise and its Control, PhD. Thesis, North Carolina State Univ., 1972, NASA CR-112314, 1972.

10. Brooks, T. F., and J. R. Bailey, Mechanisms of Aerodynamic Noise Generation in Idling Woodworking Machinery, ASME 75-DET-47, September 1975.

11. Hart, F. D., D. L. Neal, and F. O. Smetana, Industrial Noise Control: Some Case Histories, Volume 1, NASA CR-142314, 1974. (Primary source—Warnaka, G. E., H. T. Miller, and J. M. Zalos, Structural Damping as a Technique for Industrial Noise Control, AIHA J., January 1972, p. 1.)

12. Davies, H. G., and R. H. Lyon, Noise Propagation in Cellular Urban and Industrial Spaces, J. Acoust. Soc. Amer., vol. 54, no. 6, 1973, p. 1565.

13. Simpson, Myles A., Noise Barrier Design Handbook (prepared for FHA by Bolt Beranek and Newman), available from NTIS as PB-266 378, 1976.

14. Kurze, U. J., Noise Reduction by Barriers, J. Acoust. Soc. Amer., vol. 55, no. 3, 1974, p. 504.

15. Moreland, J. B., and R. S. Musa, The Performance of Acoustic Barriers, Noise Con. Eng., vol. 1, no. 2, Autumn 1973, p. 98.

16. Nordby, K. S., Measurement and Evaluation of the Insertion Loss of Panels, Noise Con. Eng., vol. 10, no. 1, January–February 1978, p. 22.

17. Lockheed Missiles and Space Co., Response of a Panel to Reverberant Acoustic Excitation, Cosmic Program Abstract, MFS-21774, 1972.

18. Dowell, E. H., Acoustoelasticity (Princeton Univ. Report AMS 1280, May 1976; NASA Grant NSG-1253), NASA CR-145110, 1977.

19. Tweed, L. W., and D. R. Tree, Three Methods for Predicting the Insertion Loss of Close-Fitting Acoustical Enclosures, Noise Con. Eng., vol. 10, no. 2, March–April 1978, p. 74.

20. Wilby, J. F., and L. D. Pope, Prediction of the Acoustic Environment in the Space Shuttle Payload Bay, AIAA Paper 79-0643, March 1979.

21. Pope, L. D., and J. F. Wilby, Band-Limited Power-Flow into Enclosures, J. Acoust. Soc. Amer., vol. 62, no. 4, 1977, p. 906.

22. Piersol, A. G., and P. E. Rentz, Experimental Studies of the Space Shuttle Payload Acoustic Environment, SAE Paper 770973, November 1977.

23. Unruh, J. F., A Finite Element Subvolume Technique for Structural-Borne Interior Noise Prediction, AIAA Paper 79-0585, March 1979.

24. Dowell, E. H., Master Plan for Prediction of Vehicle Interior Noise, AIAA Paper 79-0582, March 1979.

25. Moreland, J. D., Controlling Industrial Noise by Means of Room Boundary Absorption, Noise Con. Eng., vol. 7, no. 3, November–December 1976, p. 148.

26. Lang, J., Differences Between Acoustical Insulation Properties Measured in the Laboratory and the Results of Measurements in situ, App. Ac., vol. 5, no. 1, 1972, p. 21.

27. Levine, L. S., and J. J. DeFelice, Civil Helicopter Research Aircraft Internal Noise Prediction, NASA CR-145146, 1977.

28. Howlett, J. T., S. A. Clevenson, J. A. Rupf, and W. J. Snyder, Interior Noise Reduction in a Large Civil Helicopter, NASA TN D-8477, 1977.

29. Smith, P. W., Concepts and Applications of SEA, J. Acoust. Soc. Amer., vol. 65, supp. 1, 1979, p. S97.

30. Smith, P. W. and R. H. Lyon, Sound and Structural Vibration, NASA CR 160, 1965.

31. Lyon, R. H., What Good is SEA Anyway? Shock and Vibration Digest, vol. 2, no. 6, 1970, p. 2.

32. Ghering, W. L., *Reference Data for Acoustic Noise Control,* Ann Arbor Science Publishers, 1978.

33. Lyon, R. H., *Statistical Energy Analysis of Dynamical Systems: Theory and Applications,* MIT Press, Cambridge, MA, 1975.

34. Manning, J. E., and N. Koronaios, Experimental Study of Sound and Vibration Transmission to a Shroud-Enclosed Spacecraft, NASA CR-96144, 1968.

35. Lyon, R. H., and G. Maidanik, Review of Some Recent Research on Noise and Structural Vibration, NASA TN D-2266, 1964.

36. ANSI, American National Standard Method for the Measurement of the Real-Ear Attenuation of Ear Protectors at Threshold, ANSI Z24.22-1957.

37. ANSI, American National Standard Method for the Measurement of Real-Ear Protection of Hearing Protectors and Physical Attenuation of Earmuffs, ANSI S3.19-1974.

38. Anon. Noise Labeling Requirements, Hearing Protectors, Federal Register, vol. 42, p. 120, June 22, 1977.

39. Kroes, P., R. Fleming, and B. Lempert, List of Personal Hearing Protectors and Attenuation Data, available from NTIS as HEW Pub. No. (NIOSH) 76-120, 1975.

40. Edwards, R. G., et. al., A Field Investigation of Noise Reduction Afforded by Insert-Type Hearing Protectors (prepared for NIOSH by Watkins and Associates, Inc.), available from NTIS as DHEW (NIOSH) Pub. No. 79-115, 1979.

41. Schmidek, M. E., M. A. Layne, B. L. Lempert, and R. M. Fleming, Survey of Hearing Conservation Programs in Industry, available from NTIS as HEW Pub. No. (NIOSH) 75-178, 1975.

CHAPTER 7

Specific Noise Sources and Solutions

This chapter examines in detail several major categories of industrial noise sources and techniques for the alleviation of the noise problems they present. The selection and proportional coverage of material used here was governed primarily by the amount of NASA and related aerospace research in areas important to nonaerospace noise control engineers, namely, turbulence and rotor-stator interactions within flow ducts, blade configurations of fans and compressors, combustion noise, and jet exhaust noise. There is much theoretical and experimental analysis of these topics; where prior work is innovative and relevant to general purpose noise control, references are given. There are, however, several outstanding examples of NASA-sponsored technology that have begun to show immediate practical value; for others, prototype testing has indicated a strong potential noise reduction benefit. The most salient features of these are discussed in sufficient detail to provide a reasonable idea of their mode of operation and construction, as well as their appropriateness for certain physical and acoustical environments.

7.1 JET AND TURBULENCE NOISE

This section discusses the theory of jet noise and some of the more practical examples of suppression devices from the many techniques proposed. Valve noise is essentially caused by the generation of turbulence inside pipes after formation of a jet at a restricted orifice, and thus has many properties similar to free jets. Finally, the characterization of flame-burning profiles and their concomitant turbulent effect on the ambient atmosphere, whether in a lined combustor or in free air, remains imprecise, although aerospace investigators and commercial manufacturers of burners have made increasing efforts to improve prediction and suppression techniques.

7.1.1 Jet Noise Theory

Jet exhausts create noise in fundamentally the same manner regardless of whether they are hot or cold, located in aircraft engines, or form part of an industrial process. The high velocity gas interacts with the ambient air at rest, producing severe shearing stresses. These in turn generate a mixing region of high turbulence where individual eddies act as separate sources of noise. The resultant spectrum is typically broadband with high frequency contributions from the shearing region and low frequency components from the region of larger scale turbulence downstream of the jet (see figure 7.1).

The total sound power generated by the mixing of the jet alone, W, is normally taken as dependent on the eighth power of the exit velocity after Lighthill's theory.[1] The full expression is

$$W = \frac{k \varrho S_j v_j^8}{c^5} \; W$$

where

k is the constant of proportionality (dependent on nozzle geometry concerned)

ϱ is the jet air density, kg/m^3

S_j is the fully expanded jet area, m^2

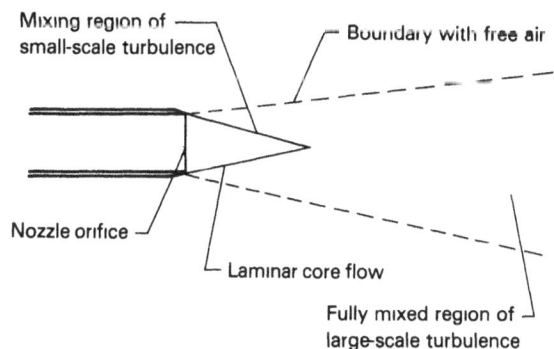

Figure 7.1 – Diagram of free jet.

93

v_j is the jet velocity, m/s
c is the speed of sound in ambient air, m/s.

The frequency at which peak sound pressure level occurs is related to the diameter of the jet nozzle and the nozzle exit velocity by the simple relation

$$f_p = \frac{0.2v_j}{D} \text{ Hz}$$

where

v_j is the jet core velocity, m/s
D is the jet nozzle diameter, m
f_p is the peak frequency, Hz.

As the ratio between the jet exit pressure and the ambient air pressure is increased, then there is a corresponding increase in the velocity of the discharge. This continues until the flow becomes sonic at a pressure ratio of around 2, when a choked condition is said to exist with little further increase in flow velocity. The parameter used to describe this is the Mach number, which is equal to the ratio of flow velocity to the ambient speed of sound propagation; sonic flow is said to exist when this ratio is equal to unity. It is not uncommon to find values of the Mach number just above this in industry when supersonic conditions are said to prevail and shock waves may be generated. Air used to pressurize industrial shop air lines can often be at a pressure more than 3.1×10^5 N/m^2 (45 psi), compared to the ambient air pressure of 1.0×10^5 N/m^2 (15 psi). Shock waves develop because the acoustic pressure disturbances can no longer radiate away from the source region faster than they are generated, thereby producing high amplitude sound waves with a characteristic sawtooth-type time history.

A prediction method has been developed by NASA for free jet noise incorporating the results of both cold- and hot-flow jets, using analytical as well as empirical data.[2] A directionality pattern resulting from convection and refraction effects is always established; figure 7.2 shows a distillation of directionality effects predicted using the above method for subsonic jets. The abscissa is defined according to

$$F = \frac{fD_e}{v_j} \left(\frac{T_j}{T_0}\right)^{0.4(1 + \cos \theta')}$$

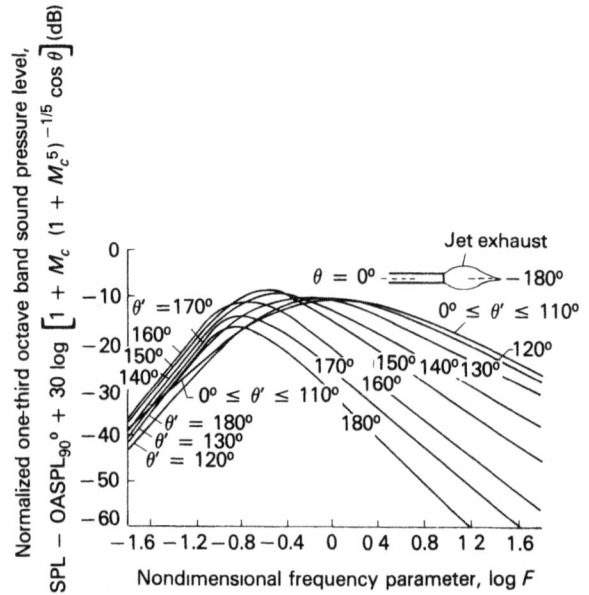

Figure 7.2 – Prediction curves for shock-free jet noise sound pressure level spectra (reference 2).

where

f is the frequency, Hz
T_j is the jet temperature, K
T_0 is the ambient temperature, K
D_e is the equivalent circular nozzle diameter, $\sqrt{4S/\pi}$, m
v_j is the jet exit velocity, m/s
S is the nozzle area, m^2

and where the effective angle θ' is given by

$$\theta' = \theta \left(\frac{v_j}{c}\right)^{0.1}$$

where

θ is the actual angle from inlet, deg
c is the velocity of sound, m/s.

The ordinate is defined as

$$\text{SPL} - \text{OASPL}_{90°} + 30 \log \left[1 + M_c \right.$$
$$\left. (1 + M_c^5)^{-1/5} \cos \theta\right]$$

where

SPL is the level in one third octave band frequency at θ
OASPL$_{90°}$ is the overall level at $\theta = 90°$.

94

Note: The convection Mach number M_c is defined as approximately 60 percent of the conventional Mach number, to take account of the average jet velocity, i.e., $M_c = 0.6v_j/c$.

Figure 7.2 includes a family of curves, each of which applies to a particular value of θ'. The effect of temperature with directional angle, as indicated by the abscissa scale, can be related to refraction effects: higher frequencies are more strongly affected. Consequently, the high frequency content increases as the angle with the jet axis increases. At $\theta' \leq 110°$ a simple model is sufficiently accurate that only one curve is necessary; in any event, at these angles the noise levels are relatively low.

In simple terms, for velocities up to the speed of sound, overall peak noise levels are of the following relative magnitudes: $+5$ dB at 150° from axis; 0 dB at 120°; and -5 dB to -10 dB at 90° to 0°. Where shock waves are generated, directionality is normally assumed to be uniform at all angles.

One study[3] indicates that the far-field acoustic intensity may be satisfactorily predicted without taking detailed measurements along the jet axis, if measurements are made near the conical outer boundary of the jet exhaust flow.

If the exhaust stream is interrupted by any kind of solid boundary, then a new source of noise is generated, the intensity of which is dependent on the location of this boundary, the geometry of the surface involved, and the acoustic nature of the surface. Subsonic jets impinging on a hard reflector may produce an increase of about 7 dB(A), and if the surface is slotted in some way then the increase may be an extra 10 dB(A) over this.[4]

7.1.2 Jet Noise Reduction

Common industrial jet noise sources include pneumatic discharge systems, steam vents, and blow-off nozzles for machine tool surfaces. Many practical procedures can attenuate jet noise output. These procedures may be categorized as core velocity reduction techniques, dissipative or absorptive devices, and the introduction of shielding secondary air flows. Reference 5 provides comprehensive performance specifications for common pneumatic exhaust and ejection silencers.

7.1.2.1 Velocity reduction

Because the exponent of the velocity term in the acoustic power relation for jets is of the eighth order, any velocity reduction has a markedly beneficial effect. Frequently, shop air pressure in compressed air lines is much higher than necessary to achieve the requisite force at the output; a reduction in pressure immediately attenuates the generated sound pressure.

One way of reducing velocity without directly reducing pressure is by stepping up the pipe diameter some distance from the orifice. To compute the total emitted sound power it is then necessary to add sound powers from the new free jet with the expanded diameter and lower velocity, as well as from the source within the pipe for which no distance attenuation should be assumed. If this increase in pipe diameter does not directly reduce the radiated noise, it may nevertheless allow the addition of in-line silencers. A similar approach has recently been evolved at the NASA Lewis Research Center for supersonic jet flow. The basic scheme is illustrated in figure 7.3, which shows the flow phenomena within the can-type suppressor.[6]

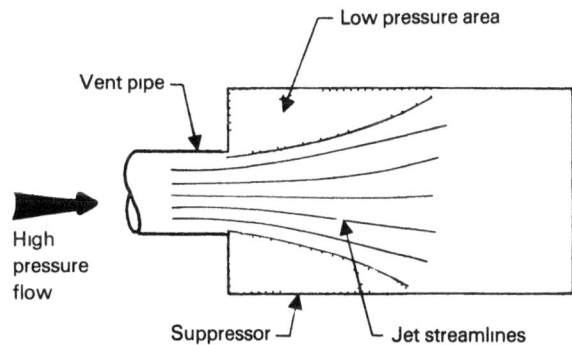

Figure 7.3 - Noise suppressor for vented high-pressure gas (reference 6).

The technique basically uses overexpansion of a supersonic jet to create a series of strong shock waves. The space between the jet streamlines and pipe enclosure is partially evacuated by the flow, thus causing the overexpansion. The gas therefore achieves a high Mach number before reaching the pipe wall, which, together with the forced turning of flow at the wall, creates shock waves in the flow. In consequence, large total pressure losses are incurred, resulting in a decrease in jet velocity and a reduced sound pressure level at the pipe exit. Acoustic tests on scale models gave maximum noise suppression up to 21 dB. The most critical feature of the design is to ensure that the overexpanded flow reaches the suppressor walls before the exit. Approximate pipe dimensions are derived from

$$\frac{L}{R_v} = 11.4 \left(\frac{R_{sp}}{R_v} - 1\right)$$

where

R_v is the vent pipe radius
R_{sp} is the suppressor radius
L is the length of suppressor.

The device is considered to be useful only where the loss of jet thrust is not important, such as in excess steam venting at power plants. Furthermore, in general only low frequency noise is suppressed.[7]

7.1.2.2 Dissipative techniques

Absorptive shrouds are usually the most effective form of jet noise suppression, because they shield against direct radiation from the source area and absorb acoustic energy as a result of multiple reflections inside the shroud. Several examples of absorptive devices are shown in figure 7.4. The introduction of diffusive elements such as "pepperpots" or grid mesh screens within the shroud breaks up the flow into many smaller jets with a lower net total acoustic power output. This also shifts the spectrum to high frequencies and may be accompanied by some net velocity reduction. Shrouds should usually be sufficiently long to prevent direct radiation at angles greater than 10° from the jet axis, and reductions of up to 20 dB are achievable. The principal defect of these devices is their propensity to generate back pressure; also, severe size constraints often apply in practical working situations.

Figure 7.4 - Examples of several kinds of jet noise silencers (by courtesy of G. W. Norris Co. and Vlier Engineering).

Many high pressure jet exhausts from compressed air machine actuators and tools have a small size; because of their size and functional design, they may be unsuited to absorptive shrouds. In consequence, it is often most convenient simply to pipe the exhaust away from the immediate working area using a thick-walled flexible tube.

One device recently developed in experimental form by NASA[8,9] uses a center body porous plug inserted along the longitudinal axis of the jet at the orifice as illustrated in figure 7.5. This is most effective for noise suppression of supersonic flows and provides a shock-free flow over a wide range of pressure ratios from 1.14 to 3.72. Shocks and the resultant noise are eliminated by pressure equalization around the jet axis. The pores between the jet flow and interior plug cavity act like settling chambers with pressures nearly equal to the ambient level, cancelling abrupt positive and negative pressure gradients within the jet stream. In addition, because of the geometrical configuration of the plug, some noise reduction of the shearing action of the jet stream with the ambient air is observed at all subsonic and supersonic flow conditions.

The most important parameters are plug length and porosity; apparently, the ratio of the plug to nozzle diameters is less significant. Supersonic flow may be present at the end of the shorter plugs tested, so that a longer configuration might be more effective at high pressure ratios. Figure 7.6 shows the extent of attenuation possible with these devices at lengths of 25 cm and 69 cm.

Porous plug suppressors concentrate noise sources in the flow, leading to a more uniform directivity pattern, and reduce the size of the large-scale structure of the flow. These results tend to enhance the performance of any acoustic baffle at the jet, further reducing noise.

Figure 7.5 – Porous plug jet noise suppressor (by courtesy of NASA Langley Research Center).

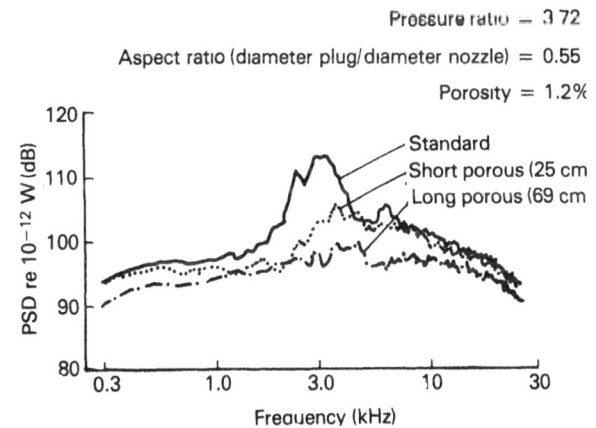

Figure 7.6 – Power spectral density of far field pressure at 90° from jet axis with porous plugs (reference 9).

7.1.2.3 Shielding by secondary air flows

Shielding by secondary air flows requires the introduction of a secondary air flow which acts as a lubricant and thereby reduces the sharpness of the shearing velocity gradient between the ambient air and jet core. Many of the techniques for achieving this have been adopted from pioneering research on aircraft jet engines.

By using a series of smaller nozzles at the exit or a suitable eductor tube, ambient air becomes

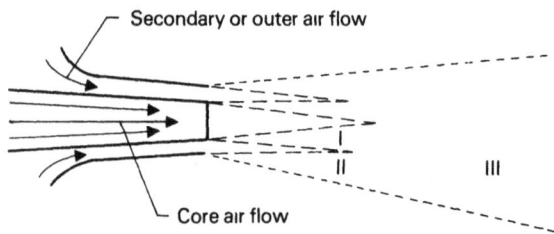

Figure 7.7 – Jet noise suppression by entrainment of ambient air flow. Noise producing regions: I, core-outer flow mixing; II, outer flow - ambient mixing; III, merged jets - ambient mixing.

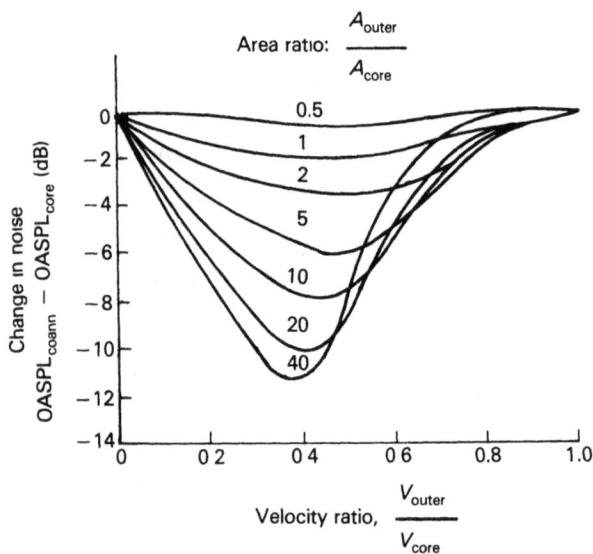

Figure 7.8 – Coannular noise reduction for jets (adapted from reference 10).

or to provide emergency pressure relief systems designed to operate either fully open or fully closed. The flow velocity of fluids and gases in pipes is usually low, but where valves are used to regulate high pressure or high velocity flows, then the ''choking'' of flow increases the velocity to the speed of ambient sound, i.e., Mach 1 (see section 7.1.1). Although there are generally no increases in velocity beyond the critical pressure ratio, the density does increase, and a mismatch in flow conditions at the vena contracta can produce downstream shockwaves (see figure 7.9).

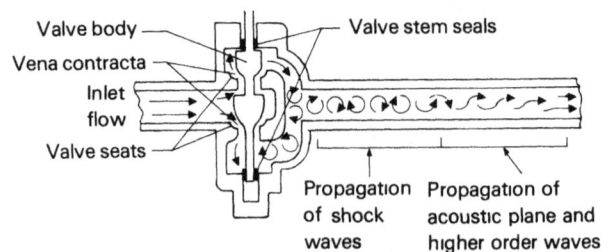

Figure 7.9 – Schematic representation of valve noise generation and propagation.

The turbulence produced by any effective valve as a by-product of the required pressure drop often leads to the generation of intense noise, with sound power levels of up to 150 dB. Valve noise is similar to jet noise, in that sound power is closely proportional to the eighth power of the orifice velocity in subcritical pressure ratio regions. In an extensive review of valve noise prediction schemes,[12] it was pointed out that for subcritical pressure ratios, the location of the most intense noise generation is about three pipe diameters downstream of the valve orifice. Above the critical pressure ratio, shock waves are present and there appears to be no simple means of accounting for their contribution to total acoustic power.

As a result of the production of shock waves and other high acoustic pressure disturbances downstream, pipe ringing may occur. In high pressure systems, the pipe walls are probably already stiff and therefore have a number of well-defined resonances easily excited by the broadband excitation forces inside the pipe. For this reason, valve noise often contains strong discrete frequency components in addition to the broad frequency spectrum normally associated with jet noise.

Noise reduction may be achieved by modifying the valve itself, silencing the sound propagated

entrained in the air flow before the jet core is fully developed, as shown in figure 7.7. The major noise sources are the core-outer flow mixing region, the outer flow-ambient air mixing region, and the merged-jets-ambient mixing region. Overall reductions of between 5 and 10 dB(A) can be achieved using this approach, and figure 7.8 shows a family of curves yielding attenuation predictions for different ratios between the outer and inner nozzle areas.

Jet noise may also be reduced by imparting a rotary component to the flow by the introduction of radial vane structures[11]; this may be particularly useful in hot gas exhausts.

7.1.3 Valve Noise

Valves are used either to control flows by creating a pressure drop, which is known as trimming, upstream or downstream of the valve with in-line units, or increasing the transmission loss of pipe walls as discussed in section 5.2.2.

Quiet valves are produced which operate according to either or both of two principles: (a) the division of the total pressure drop into a series of smaller successive pressure drop stages, each below the critical pressure ratio, and (b) the use of multiple flow restrictions in parallel, with a high length-to-diameter ratio, which cause high frictional and viscous losses. Examples of both kinds are given in figure 7.10; reductions of up to 15–20 dB are possible. Diffusers may be placed some distance downstream of the valve orifice to attenuate the noise further. Reasonable predictions of sound power level can be made for valve configurations,[12, 13] although spectral content depends heavily on individual characteristics.

For hydrodynamic flow with fluids, the major problem is to prevent cavitation at the vena contracta. Cavitation occurs because, as the fluid passes through the control valve orifice, its increased velocity causes the pressure to drop below that of the fluid vapor pressure, and hence bubbles are formed. As the flow moves downstream, the velocity decreases again with resultant pressure recovery. Implosion of the bubbles occurs when the static pressure becomes greater than the liquid vapor pressure and consequently extremely high pressure shock waves are produced. These have serious consequences, not only regarding noise, but also because of the erosion of valve parts. The most effective way of combating erosion is either

to use a series of staged pressure reductions or to utilize a cage with diametrically opposed holes in the valve trim. The flow through opposite pairs of these holes meets in the center of the cage and creates a continuous fluid cushion which prevents cavitating fluid from reaching the valve plug and seat line. Reductions of up to 6 dB are possible using this approach.

For path control of valve noise, in-line silencers are most effective when situated about two to three pipe diameters from the valve. The length of pipe between the valve and silencer should be heavily wrapped in order to reduce sound transmission through the pipe wall. Note that the outlet velocity of the silencer is controlled by the configuration of downstream pipework. For vent attenuators where gas is exhausted to open air, the diameter should be large enough to keep the merging gas velocity low (less than 50 m/s); otherwise a secondary jet noise source could be generated at the outlet.

7.1.4 Fluid Flow Problems

Hydraulic systems may present noise problems, especially where motor–pump attachments are located. The general rank order for the level of noise generated by such pumps is screw-drive, piston-drive, and gear-drive, the latter being the worst case.[14] Noise generated in the fluid line itself occurs because of sharp bends, restrictions, and internal pipe fittings, which cause cavitation and turbulence. A series of early NASA studies[15-18] examining acoustic propagation in a liquid flow line looked at the separate effects of 90° bends, closed-end side branches, dynamic response of the line, and pure tones superimposed on laminar flows of liquids with varying viscosity. It was found that the amount of energy imparted by a massive fluid to a pipe wall because of turbulent flow at internal asymmetries could be satisfactorily predicted. Some of the results showed that there was no substantial evidence for acoustic reflection effects at elbows or pipe terminations per se.

The interrelationship of the parameters influencing fluid turbulence in pipes is conveniently shown using the Reynolds number, N_R:

$$N_R = \frac{Dv\varrho}{\mu}$$

where

D is the pipe diameter, m

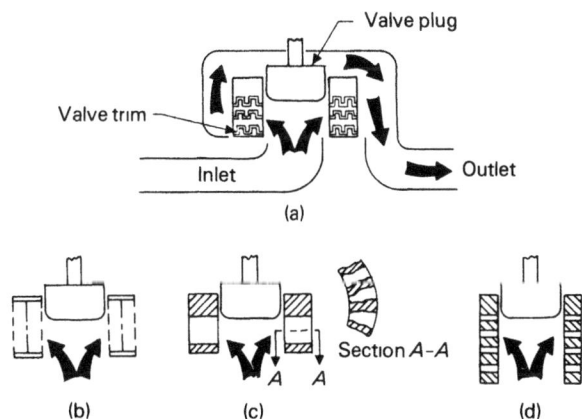

Figure 7.10 - Examples of quiet valve profiles (adapted from reference 13). (From P. N. Cheremisinoff and P. P. Cheremisinoff, Industrial Noise Control Handbook, 1977. Used with the permission of Ann Arbor Science Publishers.)

v is the fluid velocity, m/s
ϱ is the fluid density, kg/m^3
μ is the coefficient of viscosity, N·s/m^2.

When the fluid particles move parallel to each other and to the general direction of flow, laminar flow is considered to prevail. This normally occurs at Reynolds numbers of less than 1200. At higher Reynolds numbers turbulent flow becomes established, but the level of sound generated is not necessarily proportional to the value of the Reynolds number concerned.

Dissipative mufflers or tuned reactive units (see section 7.3.4) may be employed where appropriate. When a pipe size for nonturbulent flow conditions is not feasible, then acoustic energy may be attenuated using flexible hoses, surge chambers, or vibration-isolated supporting struts; in addition, the transmission loss of the pipe walls may be enhanced by wrapping or increasing the wall thickness.

7.1.5 Furnace and Combustion Noise

The noise from natural and forced draft furnaces is predominantly low to mid frequency, and one analytical study[19] confirms that combustion noise is inherently broadband with peak frequencies typically in the 400–1000 Hz range. There are also indications that the amount of heat released per unit mass of fuel is the parameter that controls the sound level at the peak frequency. Prediction

of the sound power in burners is difficult because of the wide range of variables involved, not the least of which is the flame geometry.

An examination of the characteristics of a can-type combustor with a large by-pass ratio of draft to consumed air produced the following proportional relationship for sound power level, PWL[20]:

$$PWL \propto m_f m_a^2$$

where

m_f is the fuel flow rate
m_a is the air flow rate through the can.

Oscillations and combustion roar are often both produced; typical frequency spectra for these are shown in figure 7.11.[21] The former is characterized by a discrete frequency spectrum and high levels and is not always present, whereas the latter has a relatively broadband spectrum of moderate levels and is associated with any turbulent combustion process. Oscillations in particular need to be suppressed, since they represent large-scale turbulence which can cause severe vibration and even fatigue problems in the surrounding structure.

A general discussion of combustion noise in reference 20 indicates that the generation of a roar, often exacerbated by amplification at natural frequencies of the furnace, depends on the square of the turbulence intensity. It is therefore

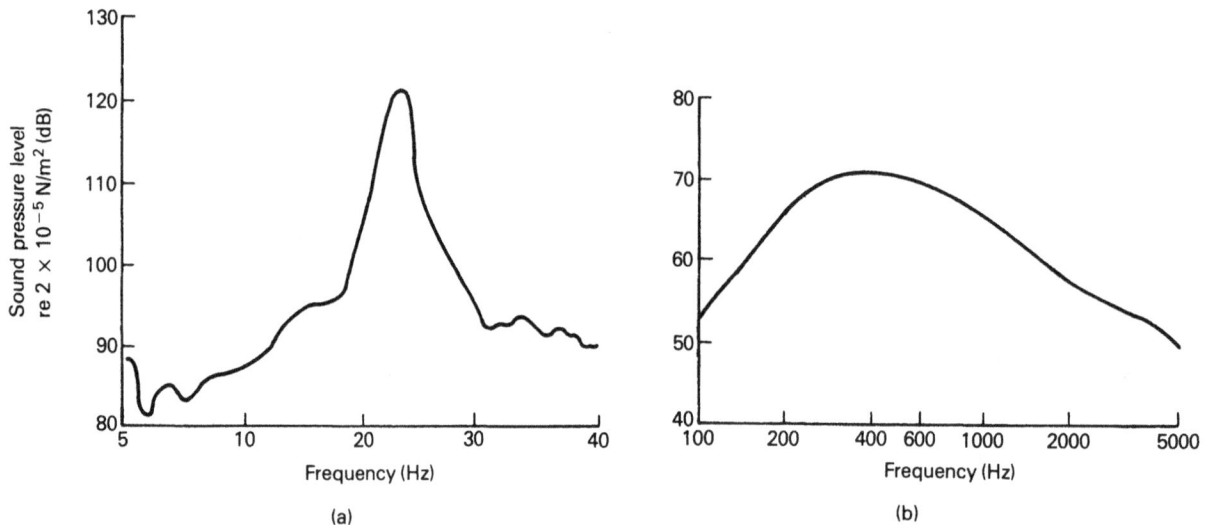

Figure 7.11 – Noise spectra for combustion phenomenon (reference 21). (a) Combustion-driven oscillation. (b) Combustion roar. (Used by permission of the Noise Control Foundation and A. A. Putnam.)

important to reduce turbulence intensity while maintaining satisfactory combustion performance. This can be achieved by reducing burner flow rates, reducing excessive swirl, and matching flow directions to the geometry of the combustion vessel. Flame-holder geometries are especially important in influencing the efficiency and roughness of combustion. Figure 7.12 shows the beneficial effect of using a plane-slotted flame holder with two counter-swirling streams over a simple perforated can holder. Apart from reducing the total power spectral density by some 15 dB, the most stable combustion is also achieved with this

configuration.[22] Other possible techniques include ducted combustors which adjust the fuel/air mixture to give shorter flames with inherently less large-scale oscillations, and the use of rods inserted into the flame region.[23] Combustion oscillations are often best eliminated or attenuated by tuned resonator cavities.

It is sometimes expedient to construct external acoustically treated plenums or enclosures and leave aside the more unpredictable approach of controlling the source noise. These can be built around a complete group of furnaces, and it is sometimes necessary to incorporate fans to provide sufficient supplemental air to feed them.

7.2 FAN AND COMPRESSOR NOISE

In a recent review of fans used in industry,[24] it was pointed out that more than half of 70 identifiable noise sources were the result of either supply and extract fans or compressor systems. These fans and compressors are in wide use for moving large quantities of air for cooling, drying, or oil-demisting operations within most manufacturing and processing operations, as well as for conventional heating, ventilation, and cooling systems for human comfort. Fortunately, relatively simple acoustic solutions may be applied to alleviate noise problems. Conventional industrial fans have lower blade loadings and lower speeds than aircraft engine fans and compressors and so they

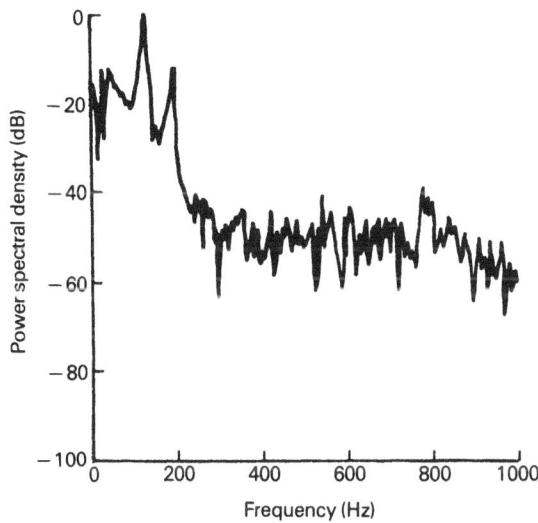

(a)

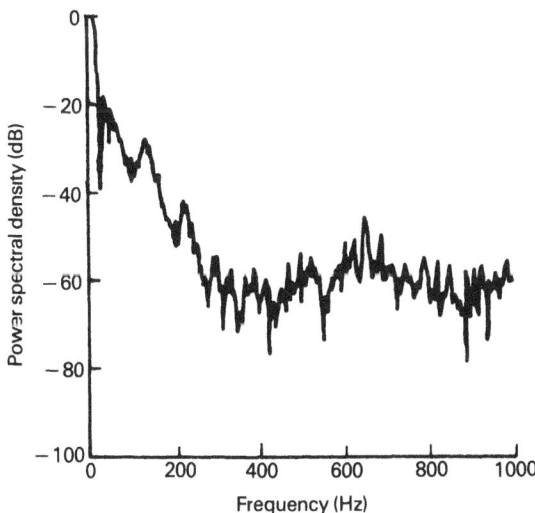

(b)

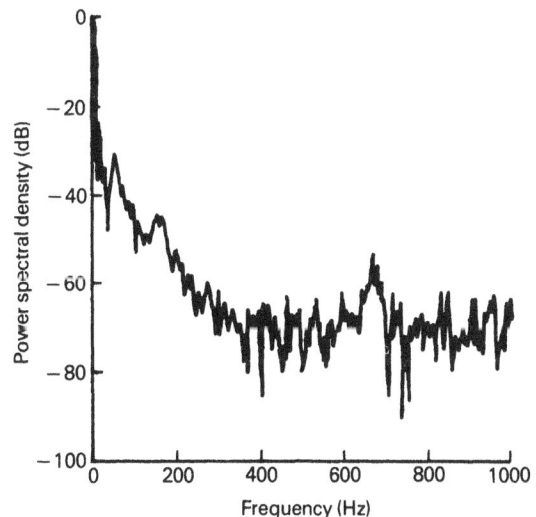

(c)

Figure 7.12 – Power spectral densities of various flame holder configurations (reference 22). (a) Plane, slotted flame holder with swirl vanes. (b) Plane, slotted flame holder. (c) Perforated can flame holder.

eliminate many of the more complex sound-generating mechanisms associated with the latter.

7.2.1 Fan Types and Noise Prediction Techniques

There are many types of fans because they have many different functions. However, once a fan has been selected for aerodynamic efficiency, load requirements, and so on, little can be done to improve source noise characteristics; noise reduction must be achieved by absorption lining and by ensuring flow stability within the vicinity of the fan unit.

7.2.1.1 Axial fans

Axial fans resemble a propeller or rotating vanes, set such that the airflow is parallel to the axis of rotation. They produce relatively little static pressure, and energy is transferred to the air by imparting a twisting motion to it. According to a detailed study,[25] the dominant noise mechanisms of low-tip-speed axial-flow fans can be separated into nonrotational and rotational noise. The former includes laminar boundary layer vortex shedding, the formation of vortices according to the clearance between blade tips and the fan housing, and blade stall, all of which generate broadband noise. Rotational noise, on the other hand, results from blade interactions with inflow distortion and turbulence, as well as with the reflected pressure field from nearby structures. This component is characterized by a series of harmonically related pure tones determined by the blade pass frequency according to

$$f_n = \frac{nKN}{60} \text{ Hz}$$

where

N is the revolutions per minute, rpm
K is the number of blades
n is the integer equal to 1, 2, 3, etc.
f_n is the nth harmonic frequency, Hz.

Axial fans are used extensively throughout industry, and a wide range of pressure ratios and air volume handling capacities are possible. The several configurations commonly available may be categorized as shrouded with guide vanes (vaneaxial); shrouded without guide vanes (tubeaxial); and freely turning without a shroud (propeller). An example of a tubeaxial fan is shown in figure 7.13.

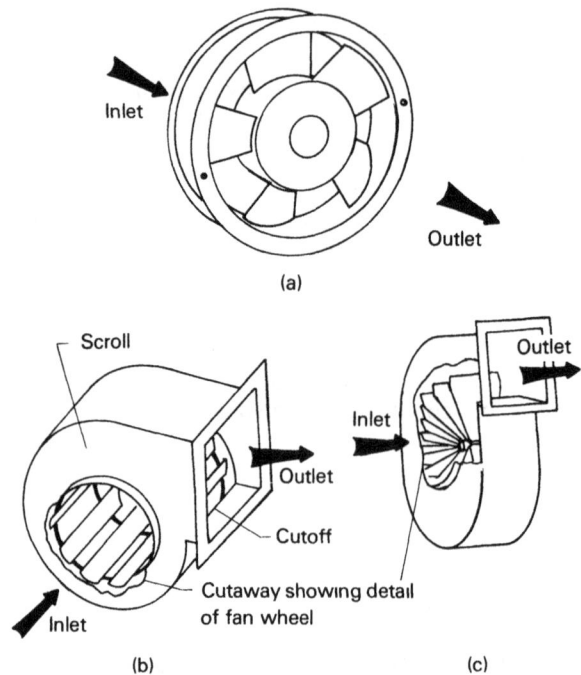

Figure 7.13 – Major fan types. (a) Tubeaxial. (b) Centrifugal tangential. (c) Centrifugal radial.

7.2.1.2 Centrifugal fans

Centrifugal fans create a pressure differential by a combination of centrifugal forces and rotating velocity. The noise generated arises from discrete tones at the blade pass frequency and from aerodynamic noise associated with the shearing action of the blades on the air flow and the resultant turbulence. There are two major types of centrifugal fans, both of which are illustrated in figure 7.13. The radial fan usually consists of fairly heavy, flat blades built around a central hub; it is of particular value in heavy-duty applications where foreign material, such as woodchippings or heavy particulate matter, passes through the unit. The blade pass frequency component is especially prominent and is calculated in the same way as for axial fans. The second type is sometimes known as the tangential fan; it consists of a circumferential wheel, within which are mounted a series of blades aligned parallel to the central inlet axis, and often with a forward- or rearward-curved characteristic with respect to the direction of rotation of the wheel. These fans tend to produce a fairly uniform spectrum, in contrast to the radial type, which have dominant low frequency components. They are used extensively in heating and ventilation systems, and the backward-curved type, in particular, has the best efficiency of all the centrifugal fans and correspondingly the lowest sound power levels.

Two parameters are of particular importance in influencing the noise characteristics of circumferential fans. The scroll around a centrifugal fan is sometimes curved so that as the angle traversed by the blades increases, so does the distance between the blade tips and the scroll casing. This helps to ensure a constant mean velocity of flow around the fan circumference, which is desirable for minimum noise production. Additionally, the clearance distance between the outlet scroll edge or cut-off and the blade tips should be around 5 to 10 percent of the fan wheel diameter. Smaller clearances than this increase the amount of noise produced.

7.2.1.3 Fan noise prediction

After the fan type is selected there are several procedures for optimizing the fan parameters to achieve maximum operating efficiency.[26, 27] However, this does not necessarily ensure the quietest running profile since blade pitch, blade curvature, number of blades, and tip solidity all act as mitigating factors, in which changes resulting in improved efficiency do not unambiguously decrease noise levels.[28] For geometrically similar fans, a flow coefficient, as defined below, exists at which a given air pumping task can be performed more quietly. There is evidence to suggest that the flow coefficient at which minimum noise occurs is unique for a particular type of fan, and it is logical therefore to compare sound power levels at these points to gain some appraisal of the value of different fan designs. Figure 7.14 illustrates this approach using the pressure coefficient as a direct measure of fan performance. At the appropriate flow coefficient, Fan A operates at peak efficiency, whereas for the same value of the flow coefficient Fan B is not operating at optimum efficiency and may well be noisier in consequence.

$$\text{Flow coefficient,} \quad \phi = \frac{3.18 \times 10^5 \cdot Q}{N D_t^3}$$

$$\text{Pressure coefficient,} \quad \psi = \frac{1000 \cdot \Delta P}{\varrho v_t^2}$$

$$\text{Efficiency,} \quad x = \frac{Q \varrho_0}{60 W_I} \frac{\Delta P}{\varrho}$$

where

Q is the air flow rate, m^3/min
N is the fan speed, rpm
D_t is the tip-tip diameter, cm
ΔP is the pressure change, kPa
ϱ is the air density, kg/m^3
v_t is the fan blade tip speed, m/s
ϱ_0 is the 1.2015 kg/m^3
W_I is the fan input power, kW.

Moreover, for a given type of fan and at the same value of the flow coefficient, effects on the sound power generated resulting from diameter and velocity changes are given by

$$\frac{W_2}{W_1} = \left(\frac{D_{T2}}{D_{T1}}\right)^2 \quad \text{and} \quad \frac{W_2}{W_1} = \left(\frac{v_{t2}}{v_{t1}}\right)^X$$

where
W_1, W_2 is the initial and final sound powers respectively, W
X is the fan velocity exponent.

Note: The velocity exponent may only be determined experimentally by measuring sound power at a large variety of velocities. It normally lies in the range 5.5–6.5 and is a function of both the flow coefficient and fan design.

Several empirical equations give a rough estimate of fan noise in regard to such parameters as static pressure developed by the unit, volume flow rate, and horsepower capacity (see, for example, reference 29). However, a more precise prediction scheme has recently taken into account in more detail the fan type and fundamental blade pass

Efficiency, x — — —
Pressure coefficient, Ψ ———

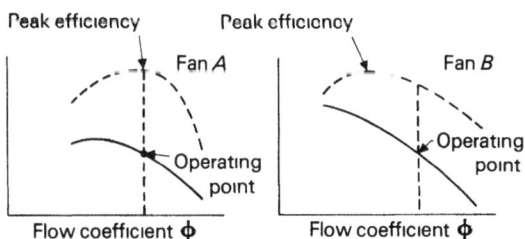

Figure 7.14 - Comparison of optimum fan system operating points (reference 26). (Used with the permission of the Noise Control Foundation and R. C. Mellin.)

frequency.[30] The basic empirical data have been reduced to specific sound power levels expressed in octave bands, as shown in table 7.1; these are the sound power levels generated by a fan type when the flow rate is 1 m³/sec and the static pressure 1 kilopascal.

To obtain the actual sound power level, PWL, the specific sound power levels for the relevant type of fan are chosen, PWL_{sp}, and adjustments made as necessary according to the following relations:

$$\Delta PWL = 10 \log Q + 10 \log P \quad dB$$
$$PWL = PWL_{sp} + \Delta$$

where

Q is the air flow rate, m³/s
P is the total pressure, kPa.

The blade pass frequency component of fan noise is included by adding the appropriate blade frequency increment, BFI (see table 7.1) to the specific sound power level of the octave band which includes the blade pass frequency, as calculated from

$$f = \frac{KN}{60} \quad Hz$$

where

K is the number of blades
N is the revolutions per minute, rpm.

Finally, for estimating the values of propagating sound pressure levels, this procedure gives the total sound power generated and it is normal to subtract 3 dB in each octave band to obtain the

Table 7.1 – Specific sound power levels for various fan types (reference 30).

Fan type	Wheel size (m)	Octave band center frequency (Hz)								BFI
		63	125	250	500	1000	2000	4000	8000	
Centrifugal, radial (material handling)	>1	98	94	90	87	83	78	75	74	7
	<1	110	106	100	91	88	85	80	86	7
Centrifugal, radial (pressure blowing)	>1	93	87	90	87	85	80	78	77	8
	>0.5, <1	103	96	96	93	93	88	86	85	8
	<0.5	111	105	106	98	92	87	86	81	8
Centrifugal, tangential (airfoil, backward curved, or backward inclined blades)	>0.75	80	80	79	77	76	71	63	55	3
	<0.75	84	86	84	82	81	76	68	60	3
Centrifugal, tangential (forward curved)	All	95	91	86	81	76	73	71	68	2
Vaneaxial	>1	87	84	86	87	85	82	80	70	6
	<1	85	87	91	91	91	89	86	80	6
Tubeaxial	>1	89	87	91	89	87	85	82	75	7
	<1	88	89	95	94	92	91	85	83	7
Propeller	All	96	93	94	92	90	90	88	86	5

BRI refers to blade frequency increment, which measures the relative magnitude of the discrete tone at the blade pass frequency.

Power levels expressed in decibels re 10^{-12} W.

sound power radiated from either the inlet or outlet alone.

Example: Calculate the sound power levels propagating upstream in octave bands for a six-bladed axial fan of tip–tip diameter 1.5 m, and turning at 1000 rpm, capable of producing an air flow rate of 4 m³/s and a total pressure of 2 kPa.

Octave Band Center Frequency (Hz)

| 63 | 125 | 250 | 500 | 1000 | 2000 | 4000 | 8000 |

from table 7.1

PWL$_{sp}$ for vaneaxial fan < 1.0 m

| 85 | 87 | 91 | 91 | 91 | 89 | 86 | 80 |

from $\Delta = 10 \log Q + 20 \log P$
and substituting, $Q = 4$ m³/s
and $P = 2$ kPa
$\Delta = 10 \log 4 + 20 \log 2 = 12$ dB

Therefore actual sound power level,
$$PWL = PWL_{sp} + 12$$

PWL under operating conditions

| 97 | 99 | 103 | 103 | 103 | 101 | 98 | 92 |

from $f_n = KN/60$
and substituting $K = 6$ and $N = 1000$ rpm
$f_1 = 100$ Hz

Blade pass frequency component, BFI

| 0 | 6 | 0 | 0 | 0 | 0 | 0 | 0 |

Total corrected

PWL

| 97 | 105 | 103 | 103 | 103 | 101 | 98 | 92 |

Upstream PWL

| 94 | 102 | 100 | 100 | 100 | 98 | 95 | 89 |

Several researchers working on comprehensive fan noise prediction techniques[31,32] have indicated that where the inflow is isotropic and homogeneous, adequate procedures already exist, but in the more general case of disturbed or turbulent nonisotropic flow there is a serious dearth of knowledge.

7.2.2 Fan Noise Reduction

After choosing the most suitable fan for a particular application from both utility and acoustic points of view, little can be done further to reduce source-generated noise. However, blade modification techniques may reduce vortex shedding and subsequent noise generation.

Vortex shedding from fan blades occurs as a result of instabilities in the laminar boundary layer on the low pressure side of the blades, which results in a very narrow frequency band of noise. This may be reduced considerably by changing the boundary layer from laminar to turbulent flow before reaching the blade trailing edge using serrations which create small-scale vortices.[33,34] The serrations break up the blade tip vortices and change the character of the fan wake vortex noise from periodic to broadband. With tip speeds less than 135 m/s, attaching serrated brass strips to the leading edge of a rotor gave an overall sound pressure level reduction of 4–8 dB on a 1.5-m-diameter rotor.[33] With proper design of the serrations, one study showed that potential reductions of up to 37 dB in generated sound power were possible.[34] Figure 7.15 shows some typical reductions in noise level with the use of various lengths of serration strips.

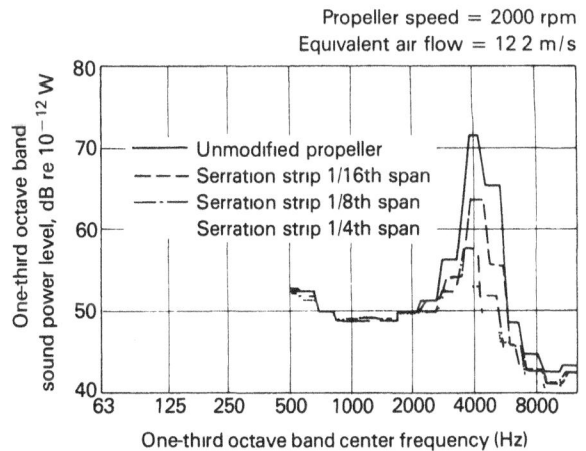

Figure 7.15 – Vortex noise reduction using serrated strip on propeller (reference 35). (Used with the permission of the Noise Control Foundation and R. E. Hayden.)

Serrations on the trailing or leading edges of low speed rotors can reduce the noise of small-scale ventilation fans, as is illustrated in figure 7.16. This design gives a reduction of 9 dB at a characteristic 2-kHz pure tone using serrated trailing edges without sacrificing fan pressure performance.

An advanced blade design concept has recently been tested by NASA in conjunction with several commercial contractors.[36] By sweeping the blade leading or trailing edges, fluctuating forces caused by a distorted inflow may to some extent be cancelled and shock formation resulting from super-

Figure 7.16 – Example of small ventilation fan with serrated trailing edge (by courtesy of Rotron, Inc.).

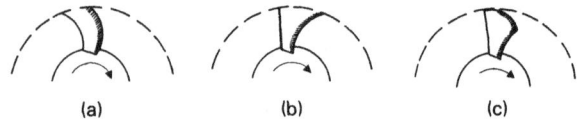

Figure 7.17 – Forward view of several swept blade configurations with subsonic leading edges (reference 36). (a) Swept back. (b) Swept forward. (c) Compound sweep.

$$\text{IL, PWL} \propto \frac{B v_{rel}^4}{L_x^2} \quad \text{dB}$$

where

B is the turbulent energy
v_{rel} is the relative flow velocity
L_x is the axial distance from fan center.

To ensure minimum turbulence and fairly uniform flow conditions at a fan inlet, there should be a distance of at least one fan diameter between a fan and any bends or upstream elements such as reduced cross sections, side branch take-offs, and so on, to allow some settling of the air flow. In this connection, all flexible connectors should be as tight as possible and tapered inlets and cross-sectional area change sections should be used. These general principles are illustrated in figure 7.18; they apply to centrifugal fans as well.

For ground static testing of aircraft engines with the kind of turbulence-free inflow typical of an in-flight situation, NASA has developed a number of acoustically transparent configurations which are fitted around the engine duct opening.[38] These inflow control devices reduce turbulence before the air reaches the fan. They are constructed of a honeycomb structure, sometimes with perforated plate panels as well. All of them were successful at reducing blade pass frequency tones and conceivably may be of value in reducing inflow turbulence in more common duct noise problems, although there may be an associated pressure loss penalty.

Having taken all practical measures to reduce fan noise at the source, absorptive parallel or co-annular baffled silencers may provide good high frequency attenuation with minimum pressure loss. It is particularly important to minimize pressure losses, since many industrial fans are of the low pressure type and are extremely susceptible to blockages in the flow on either side. This can be accomplished by using silencers with a cross-

sonic rotating speeds prevented. Sound power is proportional to the velocity component normal to the blade leading or trailing edge, f_n, according to a factor of $(v_n)^6$. By progressively sweeping a blade along its span (see figure 7.17), this component is gradually reduced. Thus by appropriately designing the sweep for the particular aerodynamic environment in which a fan will operate, a considerable reduction in vortex shedding noise may be achieved using one of the basic configurations shown in figure 7.17. Reference 36 gives a comprehensive account of design procedures for sweeping rotor blades and stator vanes with regard to aerothermo-dynamic design parameters, as well as geometric and acoustical considerations.

The effect of entry turbulence on fan blade loading is particularly important, especially because the greatest inflow turbulence occurs in the boundary layer right next to the duct wall, where the fan blade speed is highest.[31,37] The indications are that blade root and tip regions are both primary sources of sound, and that boundary layers are often present at both. One relationship evolved after a comprehensive study of turbulence noise[31] showed that both the sound intensity level and sound power level may be related to the turbulence energy as

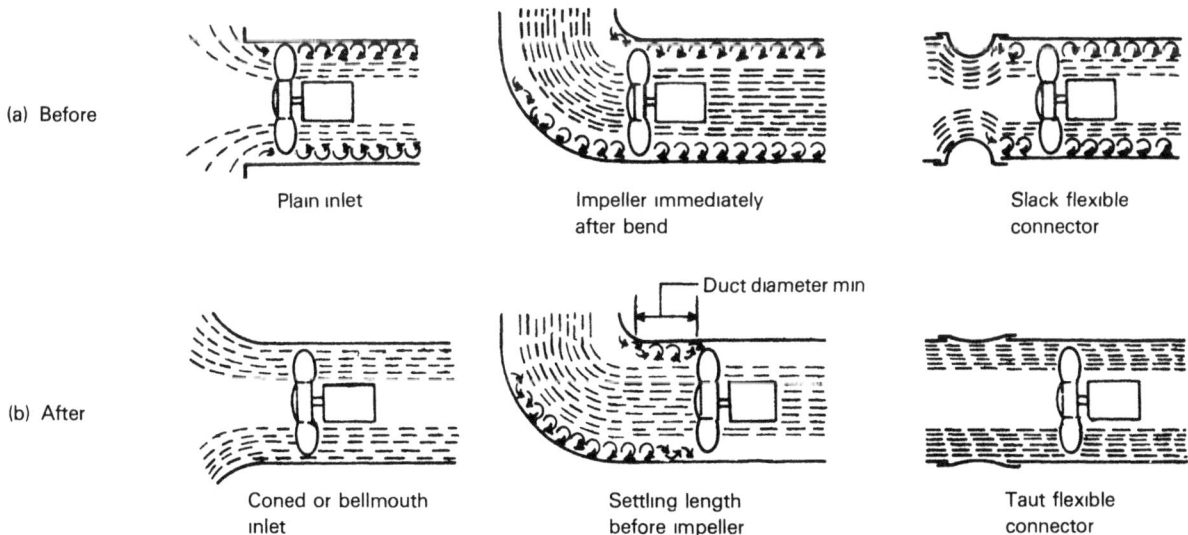

(a) Before

Plain inlet

Impeller immediately
after bend

Slack flexible
connector

Duct diameter min

(b) After

Coned or bellmouth
inlet

Settling length
before impeller

Taut flexible
connector

Figure 7.18 – Methods for reducing turbulent flow conditions at fans. (From Ian Sharland, Woods Practical Guide to Noise Control, *1971. Used by permission of Woods Acoustic.)*

sectional area of 1.25 to 1.5 times that of the fan duct, and by using turning vanes at bends. Plenums are frequently used to good effect in low pressure air heating and ventilation systems, and their attenuation properties are considerably enhanced by sound absorptive lining and baffles openings. All of these techniques are discussed in more detail in section 7.3.4.

7.2.3 Compressors and Rotor-Stator Interactions

There are a multitude of compressor configurations and power ratings. Most machines driven by reciprocating units have rotational speeds ranging from around one hundred revolutions per minute up to thousands of revolutions per minute. In industry, where compressors usually feed compressed air to power tools and blow-off jets, fairly slow speed compressors are used; these produce predominantly low frequency spectra with major contributions at harmonics of the compressor rotational speed. Larger units produce a faster rise in the work capacity than in acoustic output, and the use of larger but slower machines can provide quieter running. The noise radiated from the machine casing becomes important only when the inlet and outlet ducts are effectively silenced, but the level is governed chiefly by the piston speed and not the mechanical power produced by the machine.[39] Rotary positive displacement type blowers suffer from acoustic problems similar to those of their reciprocating cousins, although

generally their dominant frequencies are much higher. Sliding vane rotary pumps are noisier than screw types.

Compressor noise is best treated at or near the source, although, as in many other situations, it is often more straightforward and economical to use an appropriately designed acoustic enclosure. For reciprocating machines, which produce dominant low frequency sound, reactive silencers are best because dissipative silencers would need to be inordinately large to achieve the same results; conversely, for rotary machines absorptive units are usually satisfactory (further details of silencer design are given in section 7.3.4). Noise radiated from the machine casing may be reduced by damping, by adding mass, or by stiffening, depending on panel characteristics; wrapping the inlet and outlet ducts also reduces noise (see section 5.2.2). Vibration isolation of the pump, possibly using an inertia block, may be necessary if there is substantial structure-borne propagation through the mountings (see figure 5.27, above).

Mechanical noise is usually caused by worn bearings, and in this kind of application it may be best to replace ball or roller bearings with sleeve-type fittings. Suitable bearing lubrication is for maximum benefit; the general order of increasing noise level is oil-flooded rotary pumps, oil-lubricated bearings, and carbon composite bearings (oil-free).

The major source of aerodynamic noise in compressor units is from rotor-stator interaction.

Compressors usually function by drawing air past inlet guide vanes which act to steady the flow, and then compressing the flow drawn past the driven rotor against a series of static vanes or stators, which have a fixed and regular geometric relationship with respect to the rotor configuration. A schematic of a single stage unit is shown in figure 7.19; multistage units have additional rows of rotors and stators and can thereby achieve greater power efficiency and generate high pressures.

As a free rotor rotates there is no change in the blade pressure loading, but with the presence of a nearby regular obstruction such as stator vanes, pressure fluctuations are experienced because of the distorted flow. These are not abrupt as in a "slap"; rather, they arise from velocity gradients, which change the angle of attack at the blades and cause a corresponding flux in the lift forces. The magnitude of such fluctuations depends on the size, proximity, and aerodynamic influence of the obstacle. This interaction does not produce any new tones, but simply enhances the noise level output already present at the fundamental blade pass frequency and associated harmonics. This is illustrated in figure 7.20, which shows the generation of a one-lobed spinning pressure pattern because of the interaction of a four-bladed rotor with three flow guides a, b, and c. For each one-quarter turn of the rotor, a complete cycle of the interaction pressure pattern occurs; therefore, for a complete rotor cycle four interaction cycles occur, thereby enhancing the fundamental blade pass frequency (at $n = 1$) and harmonics thereof. Multilobed spinning pressure patterns excite higher harmonics of the blade pass frequency.

The parameters of interest are the number of lobes, m, and phase rotational speed of the interaction patterns, N_m, calculated as[41]

$$m = nF + kH$$

where

n is the integer number
F is the number of fan blades
H is the number of stator vanes
k is the positive or negative integer values, i.e. ± 1, ± 2, . . .

and

$$N_m = \frac{nFN}{m}$$

where N is the rotor rotational speed, rpm.

The particular blade pass frequency harmonic excited by a particular lobe pattern m is given by

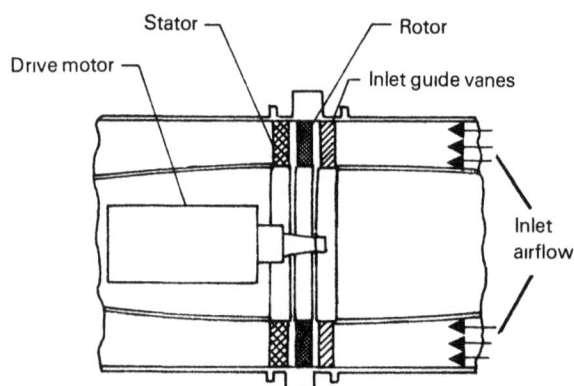

Figure 7.19 – Schematic diagram of single stage compressor.

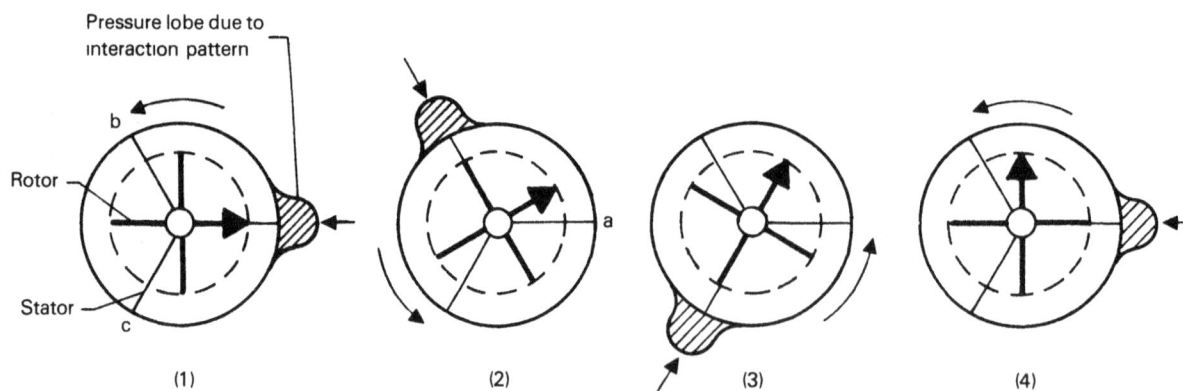

Figure 7.20 – Generation of an interaction pressure pattern (reference 40). (From Lewis H. Bell, Fundamentals of Industrial Noise Control, 1973. Used by permission of Harmony Publications.)

108

$$f_n = \frac{nFN}{60} \text{ Hz}$$

where n is the integer number 1, 2, 3, . . .

Example: In figure 7.20, the number of fan blades is 4, and the number of stator vanes is 3, thus m is calculated as

$$m = n4 + k3$$

Therefore, for the fundamental ($n = 1$),

$$
\begin{array}{lll}
\text{when} & k = -1 & m = 4 - 3 = 1 \\
& k = +1 & m = 4 + 3 = 7
\end{array}
$$

The phase rotational speed of the interaction pattern is given by

$$\text{for } m = 1 \quad N_m = \frac{4N}{1} \quad (n = 1)$$

ι.e., 4 times shaft speed as illustrated in figure 7.21;

$$\text{for } m = 7 \quad N_m = \frac{4N}{7} \quad (n = 1)$$

i.e., 4/7 times shaft speed or 1/7 phase speed of one-lobed pattern because seven times as many lobes generate the same frequency.

In designing for reduced interaction noise, it is important to avoid combinations of vane and blade configurations which produce a phase rotational speed for the pressure interaction pattern which is many times the original drive shaft speed. This is because as the speed of the pressure fluctuations approaches that of ambient sound propagation, i.e., Mach 1, there is a rapid increase to peak noise level and maximum acoustic power is radiated. Axial and centrifugal fans as well as compressors display these effects, and one of the best ways of mitigating them is to optimize the number of blades and vanes so that a high lobe number m is generated, thus reducing the radiation efficiency of the interaction because of the resultant lower phase speed, N_m.[24,40,41] Small changes in the relative twist angle between rotor and stator blades do not apparently affect the noise level, although large changes do have an influence at close rotor–stator spacings. There has also been some indication that increasing the stator chord length is not necessarily a useful expedient unless the rotor-to-stator spacing is at

several rotor chord lengths. Figure 7.21 shows the effect of increasing rotor-to-stator spacings for several experimental blade configurations. Generally, this axial separation should be of the order of at least 1.5 rotor chord widths,[41] to make a worthwhile reduction in the strength of the interaction. Another study[43] indicates that for comparable geometries and spacings, inlet guide vane–rotor interactions were much higher in level than rotor–downstream-stator interactions for single stage axial flow compressors.

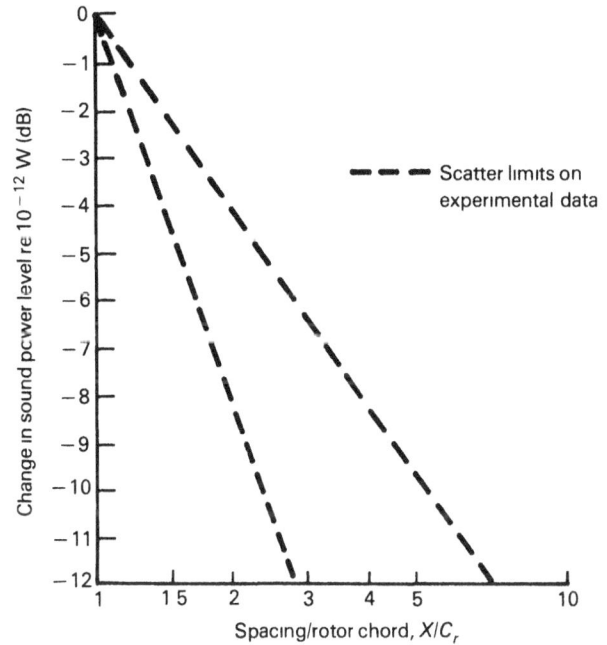

Figure 7.21 – Changes in rotor–stator interaction sound pressure levels due to rotor–stator spacing (adapted from reference 42).

7.3 DUCT-BORNE NOISE

Duct work is present in most industrial plants, either as part of the ventilation and heating system, or as an integral part of the production process, such as structures around inlet chutes and conveyors carrying components to and from the processing area. This section examines most of the aeroacoustics work performed on acoustic propagation in ducts in the presence of flow; the influence of various kinds of duct liner and means of optimizing these; and, finally, the effect of duct in-line elements such as bends, open-ends, internal vanes, and mufflers.

7.3.1 General Propagation

Unlike most other enclosures in acoustics, ducts do not generally contain diffuse sound fields since most of the acoustic energy typically propagates parallel to the duct axis. The most prominent sound waves are plane waves, which corkscrew around the duct cross section with a general direction of propagation along the duct axis. For the most part, spinning modes are only important when a driving fan or compressor, located within the duct cross section, is operated at very high speeds such that the blade tips exceed the speed of sound, and when the duct section under consideration is located fairly close to the generating source.

In an acoustically lined duct, energy is continually being extracted from the edges of plane waves, especially at the mid to lower frequencies. Because the effective speed of sound in porous materials is less at lower frequencies (see figure 5.2a, above), the effect is rather to bend the waves as shown in figure 7.22. The limitation on the extent of this attenuation at low frequencies is determined by the available thickness of material. At high frequencies, where the wavelength is of the same order or less than the duct width, most of the energy propagates down the center of the duct and so the lining has a diminished attenuation performance.[44] The amount of energy absorbed per unit length of a lined duct is given approximately by

$$\text{Attenuation,} \qquad \text{dB/m} = \alpha^{1\,4} \times \frac{L_p}{S}$$

where

L_p is the perimeter length, m
S is the cross-sectional area, m^2
α is the statistical absorption coefficient of lining.

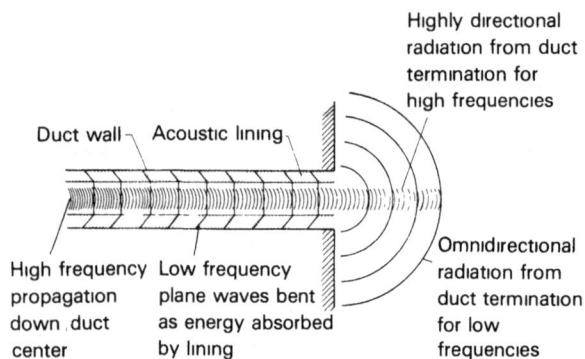

Figure 7.22 – Acoustic energy propagation in a lined duct.

This relationship indicates that maximizing the ratio of perimeter length to cross-sectional area gives the best attenuation performance.[44] Circular and square cross sections give the same order of attenuation except at low frequencies, where the extra stiffness of the circular duct reduces the degree of acoustic energy absorption and hence attenuation from resonance of the duct walls, compared to the square duct. By using rectangular ducts and making one of the lateral dimensions much larger than the other, for example, height greater than width, the potential attenuation is greatly increased. The advantages of this configuration, however, must be balanced against the disadvantages of the higher pressure drop involved with rectangular sections. Figure 7.23 shows the typical benefit from varying the relative duct wall separation in this way.

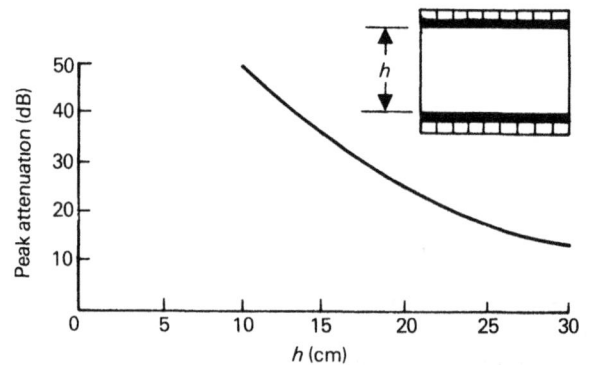

Figure 7.23 – Attenuation in ducts by varying the height (reference 46).

A variety of analytical procedures describe sound in ducts, including modal theory, purely numerical techniques, and geometric or ray tracing. Modal theory normally requires foreknowledge of the modal structure and considerable computation, and so is largely precluded from practical noise control use. Numerical techniques suffer from providing no direct notion of the mechanisms of propagation and attenuation. Ray tracing, however, although only valid at high frequencies, does at least allow the relative effects of refraction, diffraction, absorption, and reflection to be considered.

7.3.2 Internal Duct Linings

As has already been suggested in section 5.1.2.3, much attention has been given to lining

110

configurations for the relatively extreme environment of aircraft jet engines. Most of this research has focused on broadband resonator absorbers, and parameters such as acoustic treatment length and depth, core-cell dimensions, acoustic impedance, channel width, number of walls lined, air flow velocity, and number of layers in the linings, have all been examined.[45] Among the most important findings are that both the maximum attenuation and the frequency at which this occurs are strongly dependent on the duct wall separation[46]; splitter elements are most effective with an impervious central core[46]; and low flow resistance absorbent material placed in cavities behind a porous fibermetal surface gives the best overall absorption.[47]

Figure 7.24 illustrates the effect of filling 2.54-cm (1.0-in.) cavities with compressed polyurethane foam with varying overall flow resistance. The relative increase in attenuation with the number of duct wall surfaces treated is shown in figure 7.25. It is clear that this attenuation depends not only on the additional area covered, but also on the separation between these surfaces, as indicated by the poor improvement in performance obtained by treating walls C and D, which are spaced further apart than walls A and B. Another study examining the behavior of single orifice Helmholtz resonators in a grazing flow environment[48] showed that at high flow speeds, such as in many ducts, acoustic resistance is nearly linearly proportional to grazing flow speed and

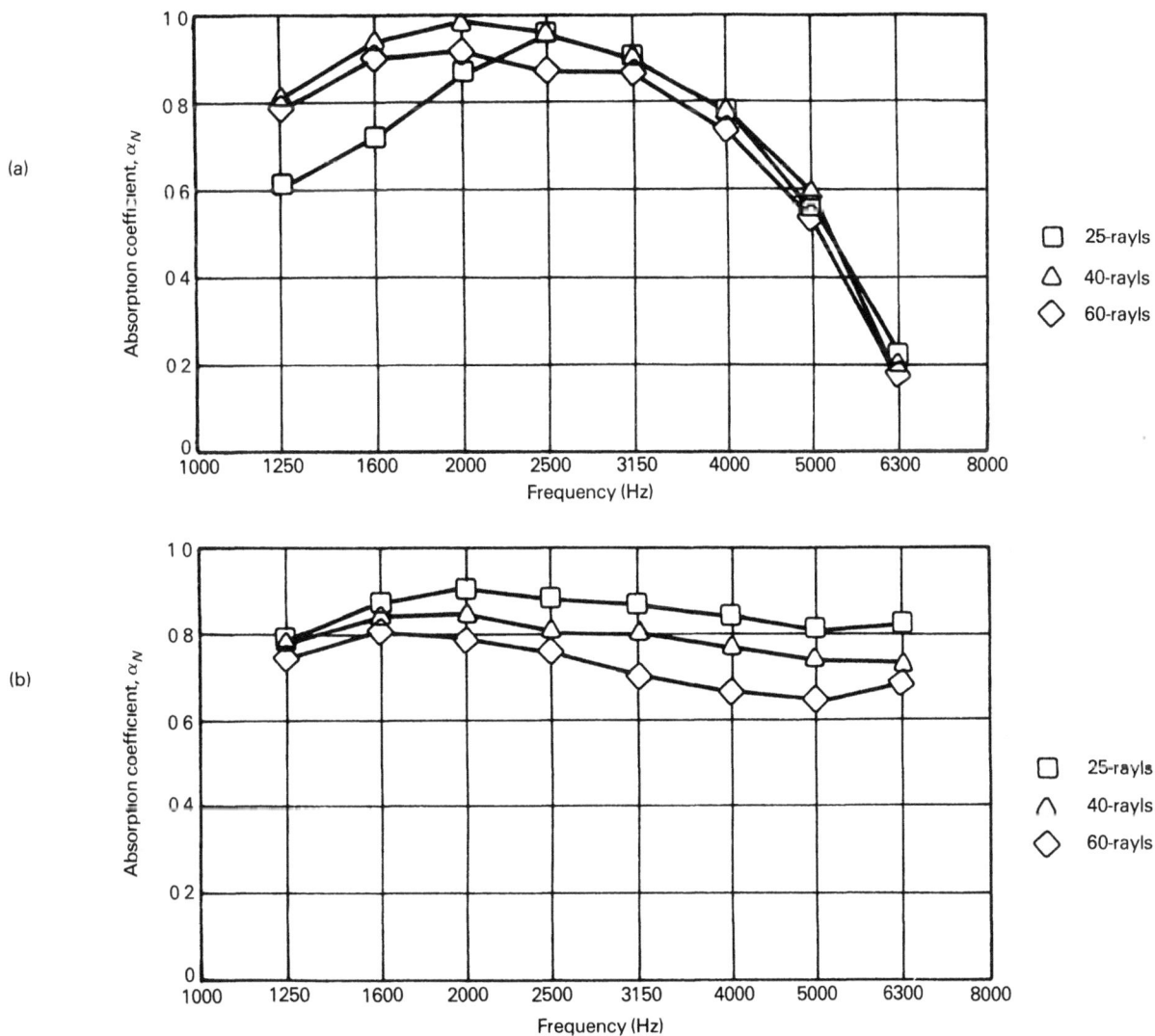

Figure 7.24 – Normal incidence absorption coefficients of fibermetal surfaces with air- and foam-filled cavities for various resistances (reference 47). (a) 2.54-cm cavity, foam filled. (b) 2.54-cm cavity, air filled.

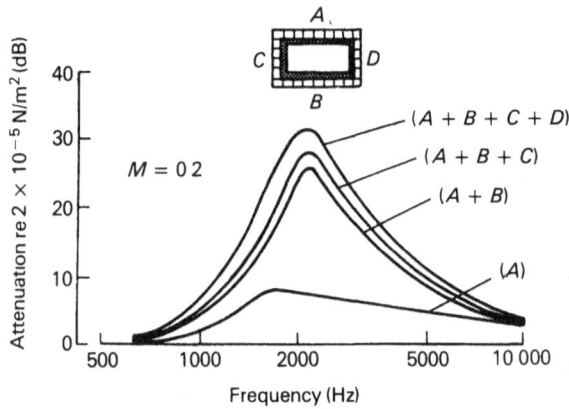

Figure 7.25 – Effect of lining various combinations of duct walls (reference 46). Attenuation values taken from duct with net flow rate of Mach number, M = 0.2.

almost independent of incident sound pressure. In fact, as the ratio of plate thickness to orifice diameter is increased, both acoustic resistance and reactance became increasingly less sensitive to grazing flow speeds.

Where spinning modes are present to any large extent, then circular ducts and noise suppressors become very sensitive to the particular modes involved, whereas annular ducts and silencers are fairly insensitive.[49] With the latter it is possible to provide reasonably effective designs for attenuating spinning modes using regular mode theory in conjunction with experimental flow duct tests; this is not the case for circular ducts, where the best approach is probably to rely mainly on empiricism. A technique for calculating the optimum wall impedance for spinning mode attenuation in soft wall ducts using as a fundamental parameter the angle of propagation within the duct with respect to the angle of incidence at the wall, is considered easier to use than conventional modal analysis techniques.[50]

A generalized approximate equation for duct lining sound attenuation, intended for the use of initial acoustic liner design, has been developed.[51] The specification of only two parameters for a particular acoustic mode, optimum impedance and maximum possible attenuation at this impedance, ΔdB_{mp} (calculated from reference 52), allows the determination of sound attenuation to be made for that acoustic mode at any selected impedance. The peak and bandwidth of sound attenuation are represented by simple ratio functions of the actual wall acoustic resistance and the

optimum resistance, so that the peak attenuation may be calculated by

$$\Delta dB_{pa} = \frac{R}{R_{mp}} \times \Delta dB_{mpa} \qquad \text{for } R < R_{mp}$$

$$\Delta dB_{pa} \cong \frac{R_{mp}}{R} \times \Delta dB_{mpa} \qquad \text{for } R > R_{mp}$$

where

R is the specific acoustic resistance

R_{mp} is the value of optimum specific acoustic resistance at frequency of peak attenuation

ΔdB_{pa} is the peak value of sound power attenuation, dB

ΔdB_{mpa} is the maximum possible attenuation at frequency for which optimum impedance is specified.

Thus for liner resistances greater than the optimum, peak attenuation is inversely proportional to the resistance, and below the optimum resistance peak attenuation is directly proportional to resistance.

The attenuation bandwidth is derived from

$$f_1/f_p \cong f_1/f_p)\text{opt}; \qquad f_2/f_p \cong f_2/f_p)\text{opt}$$
$$\text{for } R < R_{mp}$$

$$f_1/f_p \cong \frac{R_{mp}}{R} \times \frac{f_1}{f_p})\text{opt}; \qquad f_2/f_p \cong \frac{R}{R_{mp}} \times \frac{f_2}{f_p})\text{opt}$$
$$\text{for } R > R_{mp}$$

where

f_p is the frequency at which peak attenuation occurs, Hz

f_1, f_2 are the lower and upper frequencies for one-half peak attenuation at f_p, i.e., $\Delta dB = \frac{1}{2}\Delta dB_{pa}$, Hz

)opt is the frequency ratio when $R = R_{mp}$.

Thus attenuation bandwidth remains constant for linear acoustic resistances, θ, below the optimum, θ_{mp}, and increases with acoustic resistance when this is above the optimum. By the use of the following relationships, several of the more complex parameters above can be found directly:

$$f_1/f_p)\text{opt} \cong \frac{1 + 0.56\nu_p^{3/2}}{1 + 1.6\nu_p^{3/2}};$$

$$f_2/f_p)\text{opt} \cong \frac{1 + 4.21 \nu_p^2}{1 + 3.05 \nu_p^2}$$

where

ν_p is the $f_p D/c$ at the peak attenuation frequency

D is the duct diameter, m

c is the speed of sound, m/s.

These formulas may apply to more than just circular ducts, and although the equations are approximate, the unknown complete modal input and errors in analytical values of the wall impedance render superfluous any more exact analysis.

Recently, a computer program for the design of annular duct liners using perforated plate over a honeycomb structure has been developed.[53] This program is based on experimental data so that the typical effects of boundary layer turbulence, duct terminations, and so on may be included. Probably the greatest value of this program is in generating candidate designs which, for example, indicate the propitious effect of changing the number of splitter rings versus overall length of treatment.

Investigation to extend the broadband performance of resonator configurations has examined multisection liners arranged either axially or circumferentially with respect to each other. Apart from providing a composite absorbing structure with several different peak attenuation frequencies, the technique appears to enhance the overall attenuation performance by reflecting certain modes at the intraliner boundaries and redistributing others into high-order modes which are more rapidly attenuated.[54-56] Correct selection of the requisite liner impedances is imperative, and the approach seems to work best at intermediate frequency ranges. NASA has patented a device based in part on this principle which has already found commercial application[57] (see section 7.3.4). Another approach has used two layers of honeycomb cells on top of each other, but this does not appear to broaden attenuation performance as much as might be anticipated from a cursory examination.[45]

Bulk absorbers have recently undergone considerable experimental and analytical reappraisal for deployment in aircraft nacelles. The indications are that conventional multisegmenting techniques give no significant improvement, although impedance considerations are as important for optimizing suppression at discrete frequencies as they are in resonator or reactive systems.[58]

7.3.3 Effect of Lined Bends, Duct Terminations, and Splitter Elements

This section and section 7.3.4 deal with in-line attenuation elements for which only general information is provided. For more specific information on attenuation and insertion loss calculations, and specifications for individual units, the user is directed to reference 59.

Bends attenuate primarily because they reflect propagating sound waves back towards the source; this becomes important when the duct width is greater than the wavelength concerned. Therefore, low frequency sound in general is little affected by bends. A secondary phenomenon, which affects frequencies which are diffracted or reflected around the corner, is a scattering effect which converts energy hitherto propagating as plane waves into higher order transverse acoustic modes which can be attenuated much more readily. For this reason, lining bends for a short distance downstream can have a pronounced effect on the overall noise reduction achieved. Figure 7.26 shows the typical effect of lining a 90° bend both upstream and downstream for various duct widths. A resonance condition at the frequency where the wavelength is twice the duct width for unlined ducts is responsible for the peak attenuation value associated with each curve; generally, as the internal width is reduced, the low frequency attenuation performance is also reduced.

The use of sharp bends with associated absorption lining is limited by the amount of pressure drop permissible across the unit. Long radius bends or mitred bends (90°) with long internal turning vanes are excellent for minimizing pressure losses but unfortunately do not provide much attenuation. In a wind tunnel noise study[60] with high velocity flow rates, considerable attention was given to the influence of turning vanes. It was noted that the short-chord vanes turned sound waves with wavelengths less than the vane chords through an angle dependent on the incident angle at the vane row, as shown in figure 7.27. Figure 7.27 also shows the extent and configuration of wall lining treatment used to absorb sound energy at the bend; at this particular corner, the bend lining was rendered fairly ineffective at high frequencies because the vanes turned the propagating sound past the lining. One way around this problem, other than increasing the lining length downstream, is to acoustically treat the vanes themselves, although this is usually an expensive

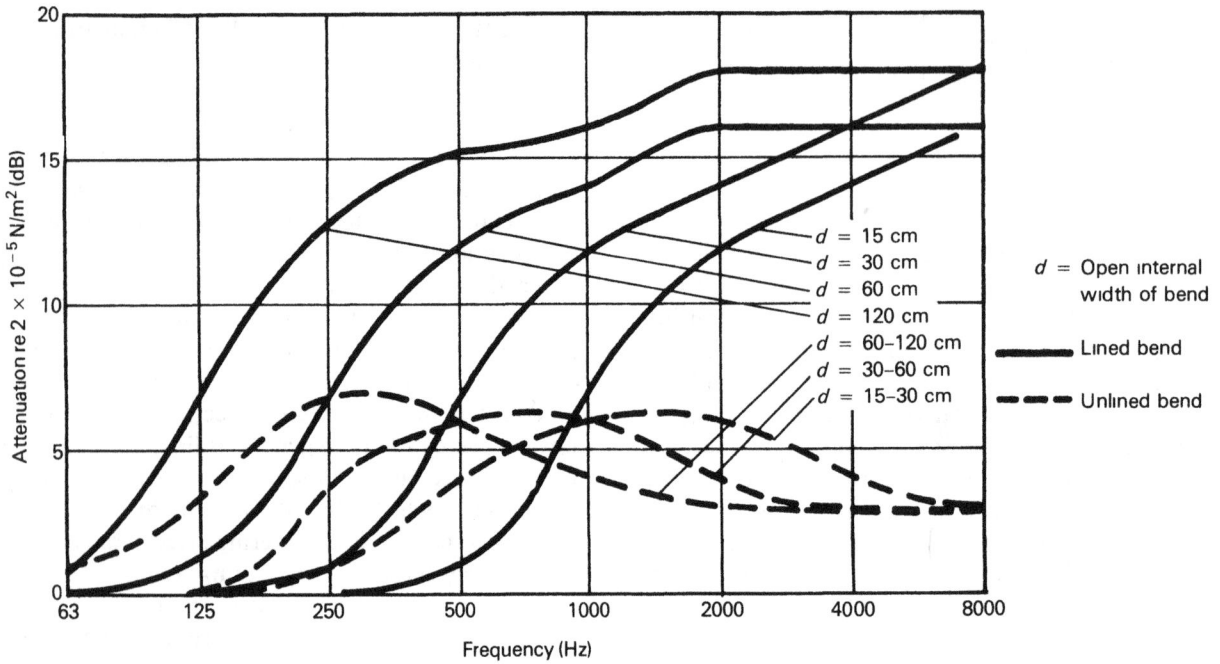

Figure 7.26 – Typical effect of acoustically lining 90° bends (reference 59).

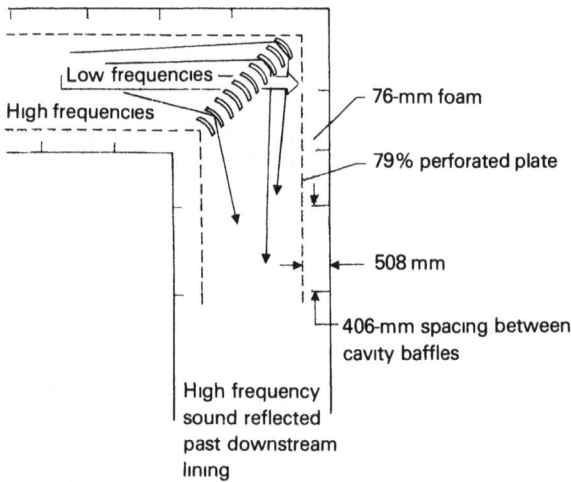

Figure 7.27 – Installation of corner vanes at an acoustically lined bend (adapted from reference 60).

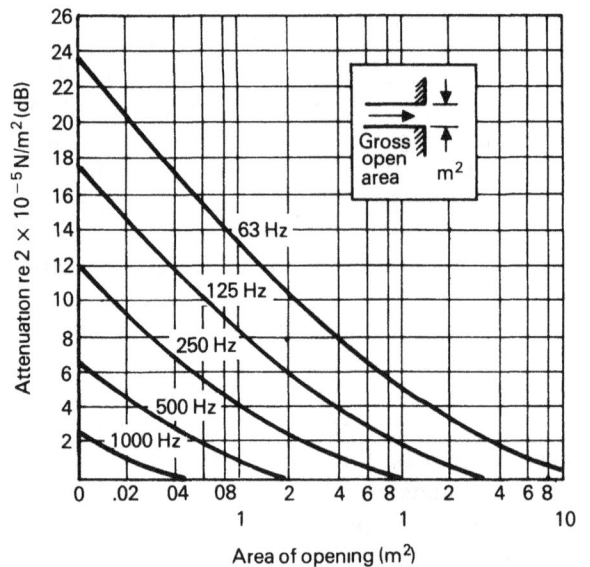

Figure 7.28 – Attenuation due to reflection at a duct outlet. (From Ian Sharland, Woods Practical Guide to Noise Control, *1971. Used by permission of Woods Acoustics.)*

expedient. The cavity baffles spaced at 406-mm intervals prevent sound from propagating along the sides behind the foam.

At duct terminations there is a large change of acoustic impedance, because elements of air through which sound energy propagates move against a much larger mass of air outside the duct. The result is an end reflection dependent on the frequency and duct area concerned. Figure 7.28

shows the degree of attenuation which occurs with various frequencies at a duct opening flush with the surrounding surface. Low frequencies do not propagate well into an open space, although the energy that does so radiates fairly uniformly from

114

the duct opening compared to the more directional high frequencies (see also figure 7.23). The effect of grilles, diffusers, louvres, and so on at the termination must be separately computed.

A much more sophisticated and exact prediction technique has been developed for the radiation of sound from a rotating blade source within circular hard-wall ducts.[61] This may be of value in appraising the validity of other more approximate approaches.

Attenuation predictions at acoustic branches which occur at angles relatively oblique to the main flow direction are based on the approximation that the acoustic energy is divided according to the relative cross-sectional areas:

$$\text{Branch attenuation} = 10 \log \left[\frac{\text{branch duct area}}{\text{main duct area}} \right] \text{ dB}$$

7.3.4 Mufflers and Duct Silencers

Mufflers and duct silencers use absorption or the reflection and expansion of sound waves due to sharp impedance changes at cross-sectional area changes and turns. The former technique relies on dissipative mechanisms and the latter on reactive mechanisms; in many silencer units both principles operate to some extent. Mufflers or silencers are used to reduce noise transmission along ducts or pipework, and the noise may be flow-generated or may emanate from a machine source. In choosing an appropriate device, particular regard should be given to the allowable pressure drop, maximum length needed, gas or fluid flow rate, and whether acoustic propagation is downstream or upstream with respect to the general flow.

7.3.4.1 Resistive units

By splitting the duct width into smaller sections with reduced width, low frequency performance may be significantly improved. The partitions for this are usually of perforated plate backed by an absorbent material such as fiber glass or foam. It is usually most convenient to use proprietary splitting units with subsection widths down to as low as 50 mm. These units can often be most satisfactorily deployed at bends, where the total attenuation is given by the total equivalent straight length of splitter plus a bend attenuation factor.

Attenuation may generally be increased by reducing the airway width and increasing the total splitter length, but these must be balanced against

the requirements for a minimum pressure loss achieved by increasing the height and total width of the complete splitter unit. In addition, the leading and trailing edges of the splitter modules should be aerodynamically designed to reduce flow resistance. The manufacturer's data should specify the acoustic performance for various rates of air flow: in general, attenuation decreases as the flow rate in the direction of sound propagation increases. Annular modules in cylindrical ducts operate on the same principles. Illustrations of both cylindrical and annular silencers are shown in figure 7.29, with attenuation graphs showing the effect of different lining configurations. An example of a large scale annular silencer for use in attenuating a 126 dB(A) level at a wind tunnel ejector by some 26 dB(A) is shown in figure 7.30.

Rather than using a fibrous backing for the acoustic baffles in the splitter units, it may be more expedient to use a resonator-type design with relatively large air spaces filled with absorber behind a perforated plate. In a comprehensive series of tests,[62] there were indications that a sawtooth type design with a 0.3 percent perforate partition is particularly useful at improving low frequency performance. Figure 7.31 shows the configurations and performance data obtained for several configurations. The use of a central partition inclined at an angle to the longitudinal axis of the baffle is intended to reduce cavity whistling caused by high speed air flows at grazing incidence, by detuning the cavity because of its varying depth. Transverse partitions across the baffles enhance absorption at the frequencies whose half-wavelength corresponds to the distance between them. Reference 62 also includes a discussion of various techniques for reducing the self-generating noise of mufflers. Some of these are illustrated in figure 7.32. Resonant tones produced by single orifice Helmholtz resonators because of the shearing action of the grazing flow layer have been shown to occur at a grazing flow speed, v_∞, given by[47]

$$v_\infty = f_{\text{res}} D/0.26 \quad \text{m/s}$$

where

f_{res} is the classical Helmholtz resonant frequency (see section 5.1.2.3), Hz

D is the orifice diameter, m.

Other techniques for improving the performance of lined baffles in splitter units include blocking the direct line of sight through the unit and the

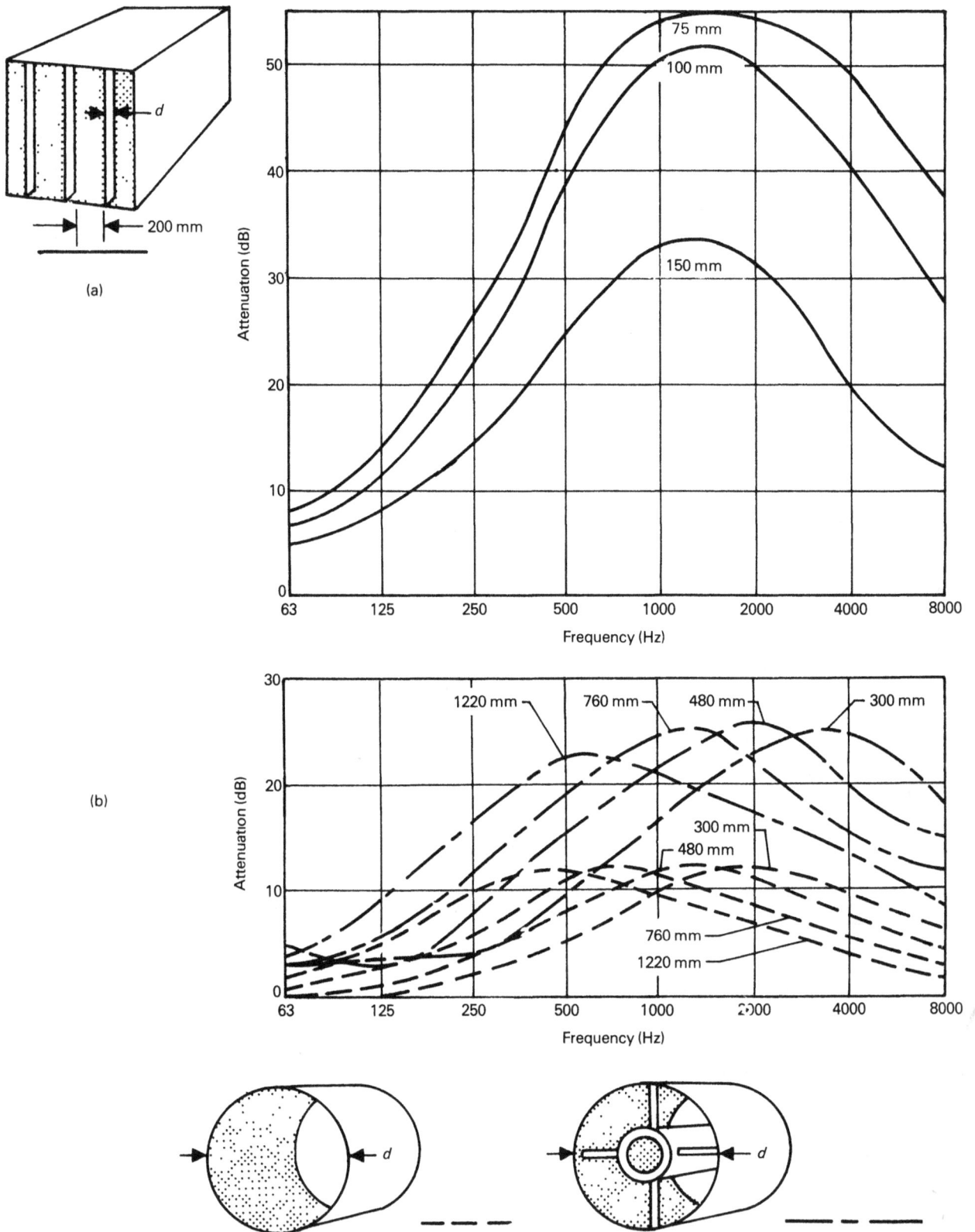

Figure 7.29 – Typical attenuation characteristics of rectangular and circular splitter modules. (a) Rectangular module. (b) Circular modules. (Adapted by Ian Sharland, Woods Practical Guide to Noise Control, *1971. Used by permission of Woods Acoustics.)*

Figure 7.30 - *Photograph of large annular duct silencer (by courtesy of NASA Langley Research Center).*

use of different liners in each separate airway.[63] The latter technique can introduce a relative phase lag between propagating components in each airway with resultant reflection and interference effects at low frequencies.

Acoustically lined plenum chambers represent a very efficient way of achieving noise reduc-

tion; their performance is enhanced by incorporating an absorbent-lined partial barrier which serves to block the direct line of sight across the chamber. The most important design characteristics are the absorption of the lining and the distance between the inlet and outlet ducts. The sound fields associated with plenums may be regarded as similar to those inside large acoustical enclosures. Figure 7.33 shows a single chamber plenum installation, and the insertion loss of this after taking account of reflection at the inlet duct is given by[29]

$$\text{InL} = 10 \log S_p \left[\frac{\cos \theta}{2\pi d^2} + \frac{1}{R_{cp}} \right]$$

where

R_{cp} is the room constant of the plenum, equal to $\dfrac{S_p \bar{\alpha}_p}{(1 - \bar{\alpha}_p)}$, m^2

d is the direct distance between inlet and outlet openings, m

θ is the slant angle between longitudinal axes of inlet and outlet ducts

S_p is the outlet cross-sectional area, m^2.

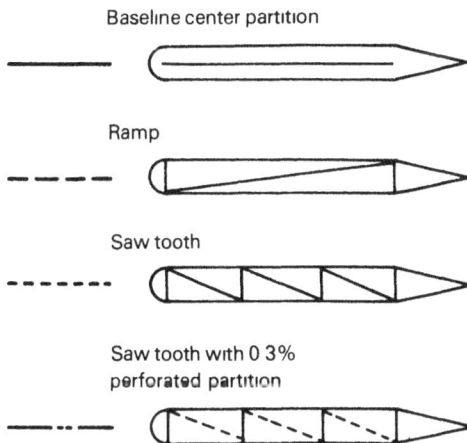

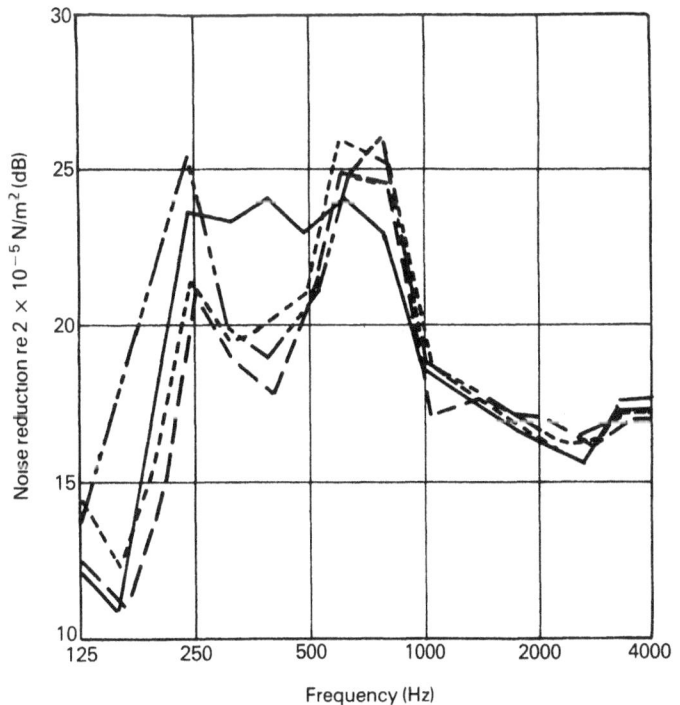

Figure 7.31 - *Attenuation performance of various staggered central partitions in acoustic baffles (reference 60).*

117

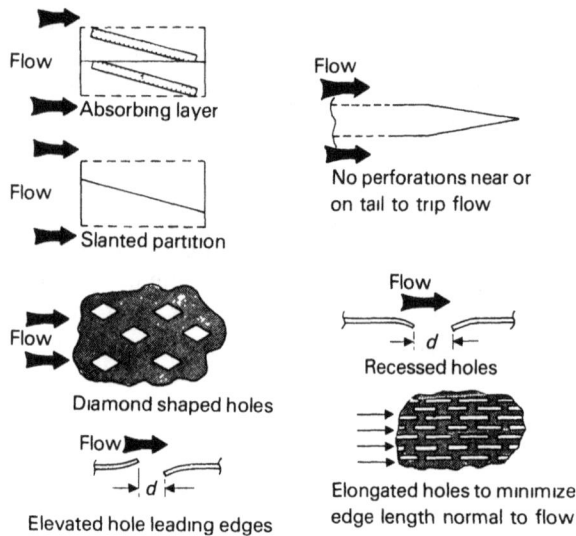

Figure 7.32 – Candidate perforated plate designs for minimizing the generation of self-noise (reference 60).

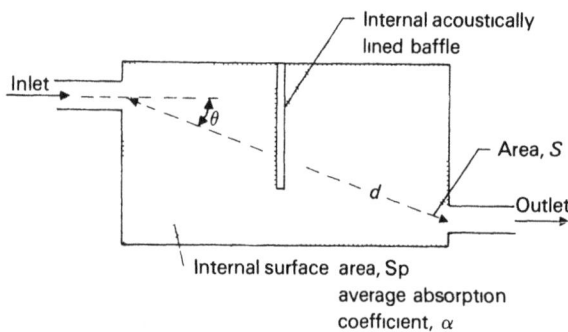

Figure 7.33 – Acoustically lined plenum chamber.

7.3.4.2 Reactive units

The most simple kind of reactive silencer is designed by changing the cross-sectional area of a short length of duct. The insertion loss is given by

$$InL = 10 \log \left[1 + \frac{1}{4}\left(M - \frac{1}{M}\right)\sin^2 \frac{2\pi L}{\lambda} \right] \text{ dB}$$

where

M is the S_2/S_1, cross-sectional area ratio

L is the muffler length, m

λ is the wavelength of frequency of interest, m.

Since the attenuation is obviously greatest when $\sin(2\pi L/\lambda)$ equals 1, and this occurs at $\pi/2$, $3\pi/2$, $5\pi/2 \ldots$, the length of silencer L corresponding to maximum attenuation is given by

$$\frac{2\pi L}{\lambda} = \frac{n\pi}{2}$$

where $n = 1, 3, 5, \ldots$

so that peak attenuation occurs at $L = n\lambda/4$.

These silencers are therefore often referred to as quarter-wave mufflers, and several of them can be placed in series to provide a broader range of attenuation. Typically their performance is best at low frequencies; a variation on this principle, the Helmholtz resonator, may be used when it is more convenient to have the muffler element as a side-branch to prevent flow obstruction (see section 5.1.2.3).

There are many prediction techniques for muffler design, several of which have been computerized. One in particular[64] improves on standard transmission line theory by accounting for the effect of mean gas exhaust flow on the overall attenuation properties of a three-chamber muffler. This NASA program has been used to design a unit capable of providing up to 11 dB of extra attenuation over standard mufflers in an operational helicopter, without any deterioration in engine performance from back pressure. The program is based on empirical data for exhaust flows in the Mach number range of 0.05 to 0.15 and expansion ratios less than 0.6.

Another example of the relevance of aeroacoustics technology to general noise control problems is the recent development to prototype stage by a commercial organization of a multiple-segmented acoustic liner in a standard truck muffler geometry (see figure 7.34). The device has a uniform inner diameter such that the direct exhaust air flow is not blocked, and it utilizes a unique combination of absorbing elements working in conjunction with reflecting elements in order to provide maximum attenuation.[56] It achieves this by forming an acoustic trap where the reflective elements on the ends direct sound energy into the central sound dissipating units. This particular application uses three cylindrically stacked segments in the center, each of different porosity and of varying depth. It provides a noise attenuation of peak levels of twice that possible with more conventional designs.[65]

An example of a serious aerodynamic noise problem is included here and the various approaches taken toward its solution highlighted.[66]

Figure 7.35 shows a wind tunnel configuration complete with installed acoustical treatment.

118

Figure 7.34 – Truck muffler utilizing multisegmented liner (by courtesy of Donaldson Company).

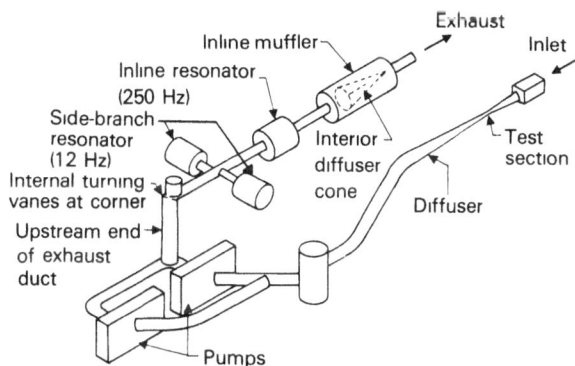

Figure 7.35 – Acoustical treatment of wind tunnel (reference 66).

The total air pump handling rate is 500 m³/min, and the inlet and discharge lines are of 35-mm steel, up to 1 m in diameter and bolted directly to the pumps. The generated spectrum contained many harmonics with a fundamental frequency of 12 Hz. The noise and vibration generated were so great that the facility could only be run for a few minutes. Measured exhaust levels were in excess of 120 dB with a maximum peak at around 12 Hz. The following procedures were implemented in order to attenuate the noise:

1. Incorporation of two reactive Helmholtz resonators with internal lining, tuned to 12 Hz.

2. Incorporation of reactive resonator, tuned to 250 Hz, for attenuating high frequency sound.

3. Design of in-line dissipative muffler and the incorporation of diffuser cones to inhibit the development of standing waves in the exhaust duct.

4. External wrapping of inlet duct with absorbent materials on high flow section and internal lining in low flow section.

Reactive techniques were mainly employed because the high pressure losses associated with dissipative muffler units would have compromised operating ability.

7.4 TRANSMISSION AND GEAR NOISE

Gear noise requires careful analysis to identify major sources and to rank-order their levels. This analysis generally requires narrow-band frequency analysis and a detailed knowledge of the mechanism of the gear-train involved. The systematic approach adopted should include an appraisal of basic gear types, major noise generating mechanisms, and their spectral character, and a detailed comparison of noise reduction measures. Unfortunately, it is rare for only single gear trains to be involved, access may be limited, and several of the major noise mechanisms may exist simultaneously, making it difficult to discover each individual contribution.

There are a number of techniques for reducing gear noise without source modification. These are discussed later in section 7.4.4 and include using completely different means of transferring and controlling mechanical power, as well as traditional path noise control procedures.

7.4.1 Gear Train Parameters of Importance

Several basic components and gear wheel parameters are shown in figure 7.36. The core of the wheel is known as the hub and is an integral part of the drive shaft. The web is the flange concentric with the shaft, and the rim is the section of the flange which incorporates the entire gear tooth profile. The pitch circle is one of the most fundamental gear characteristics, and this normally runs through points midway between the top and bottom of each tooth; pitch circles of meshing gears are always tangential to each other. The circular pitch is measured along the pitch circle and is the distance between congruent points on adjacent teeth, and the backlash is the difference between engaging tooth width and receiving tooth space width, along the mutual traverse of adjacent tooth pitch circles.

The computation of gear clash frequencies is simply made as

$$f_n = \frac{nKN}{60} \ \ \text{Hz}$$

where

n is the integer representing the various harmonics possible

K is the number of gear teeth

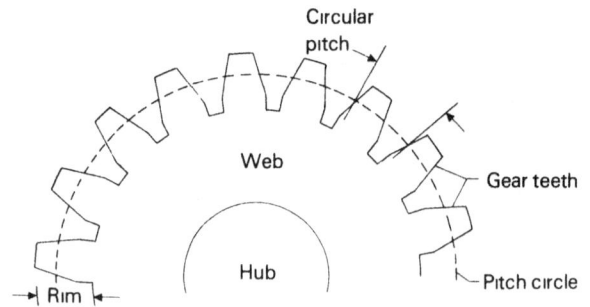

Figure 7.36 – Schematic diagram of major gear parts and important parameters.

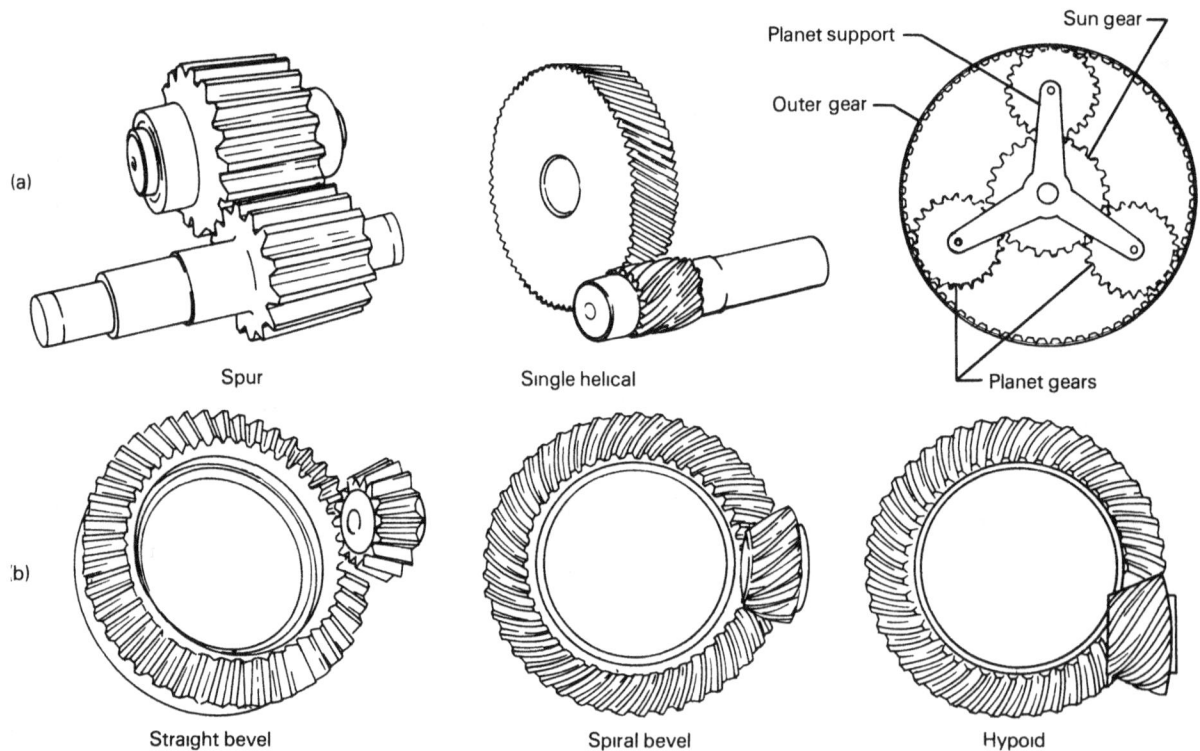

Figure 7.37 – Common gear configurations. (a) Parallel gears. (b) Nonparallel gears. (Adapted and used by permission of Machine Design.)

120

N is the shaft rotational speed, rpm

f_n is the nth harmonic frequency, Hz.

For systems of gears with a central sun gear surrounded by a series of equally sized and equally spaced planetary gears, the following relationship is used[40]:

$$f_n = \frac{nK_s(N_s - N_{ps})}{60} \text{ Hz}$$

where

K_s is the number of teeth on sun gear

N_s is the rotational speed of sun gear, rpm

N_{ps} is the rotational speed of planetary system, rpm.

Figure 7.37 shows common gear types.

7.4.2 Gear Noise Mechanisms

There are three aspects of gear quality that radically affect subsequent noise levels regardless of the precise mechanism involved. Errors in the tooth profile shape and roughness of the machined surfaces both tend to create high levels at the upper frequencies, even at low peripheral speeds, while basic pitch error is the third factor of importance.[67] Wheel quality is therefore a sensitive variable. The most significant sources of gear-meshing noise are as follows[40]:

1. Engagement—The teeth of both engaging gears are deflected tangentially, then return to their former positions after disengagement.

2. Pitch-circle impulse—Deflections of teeth during engagement mean that a continual rolling contact can never be ensured, and the friction from the resultant sliding action creates periodic noise.

3. Air-oil jet noise—This occurs when the pressure between teeth compresses and expels air and oil; shock waves may form with high-speed gears because of the high velocities involved.

4. Structural vibration—As a result of the three processes outlined above, which all directly create acoustical energy at the gear mesh frequency, radiation of acoustical energy by the wheels themselves also occurs; the extent of this radiation depends on the relationship of excitation frequencies to gear wheel natural frequencies.

A typical spectrum consists of a dominant fundamental component corresponding to the blade pass frequency and decreasing amplitudes for harmonics of this. Deviations from this pattern suggest structural resonances of some sort. Other features include side band tones generated around gear mesh frequencies because of modulation by eccentricities arising with rotation of the shaft, and harmonics at shaft rotational speeds resulting from chipped or missing teeth. Several of these phenomena are illustrated in the spectrum shown in figure 7.38.

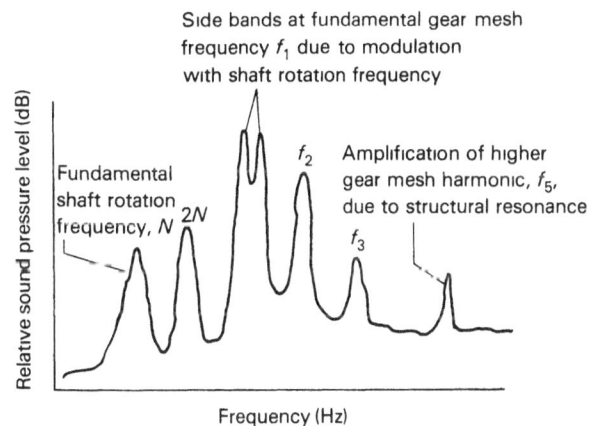

Figure 7.38 – Narrow band spectral analysis showing discrete tones.

7.4.3 Gear Noise Reduction

Although there are many possibilities for noise attenuation at source, most procedures result in very limited reduction, and an unsystematic ad hoc approach produces very poor return for the gear redesign involved.

Generally, helical gears are quieter than identical spur gears because the loading is better distributed and the impact forces reduced.[68] Worm gears are even more effective, because the continuous sliding action minimizes impulses and corresponding tooth deflections. By the same token, spiral bevel and hypoid gears can be some 5 dB quieter than their straight-bevel analogs. Aside from changing the gear type, better noise performance may be achieved by optimizing pitch-contact ratios, teeth profiles, and backlash. The use of a higher pitch contact ratio, that is, increasing the number of gear teeth, gives quieter running because the impulse loading on each tooth is lower. For the same drive speed, however, this raises the frequency spectrum, which may offset any potential benefit. Ideally, profiles should be designed so that maximum tooth contact occurs during the phase when the point of contact is inside the pitch circle, and always so that a rolling as opposed to a sliding action is favored. Backlash must be balanced between excess noise produced with a loose mesh and the mechanical drag and wear of a tight mesh.

Other than careful redesign of gear parameters, a number of secondary techniques may be used. Lubrication of all running parts and meshing surfaces is essential and can provide as much noise reduction as the approaches already discussed. By separating surfaces and providing a mechanical impedance mismatch, the oil reduces impact forces directly. A high damping factor may be obtained with synthetic gear wheel materials such as nylon, subject to load and temperature constraints. These can provide as much as 10–20 dB attenuation and have found widespread application in machines ranging from spinning looms to household washing machines. External damping can be useful where the rim and web flange is very narrow and therefore susceptible to induced resonance because of its low stiffness.

Finally, but perhaps most importantly, the advantages of total gearbox enclosure should be mentioned. This is a particularly good expedient since it involves no costly analysis and lengthy redesign programs. Adherence to the general principles of enclosure construction outlined in section 6.2.1.2 can yield excellent results. It is sufficient to reiterate here that accessibility or visibility must be maintained, removable panels should be well sealed and should have transmission loss values of the same order as the main body of the enclosure, and resonance should be avoided by stiffening and damping techniques.

7.4.4 Alternative Means of Transmission

Conventional toothed-gear drives may be replaced in several ways. Perhaps one of the most common is the use of roller-belt systems where a series of parallel roller pairs with correctly sized diameters are interconnected by flexible rubber or synthetic belts, which may be ribbed for a better grip. Unfortunately, substantial slip at high speeds or high torque ratios may occur because of stretch-ing of the belt, lack of sufficient contact resistance, and so on. This type of installation is therefore generally unsuited to systems which require a high loading or exact speed transformation.

Another well-known technique is to use traction element drives, although this technique has serious constraints. Traction element drives offer quiet and smooth power transmission, whereas even with perfectly machined gear teeth there are always torsional fluctuations caused by load transfers between teeth. Recently, however, with the active participation of the inventor of a new traction drive geometry, the NASA Lewis Research Center has developed and demonstrated considerable advantages in using this technology in a much wider range of situations than was hitherto possible.[69]

The new approach uses a number of planetary rows of stepped rollers, normally two or three, to achieve power transfer between concentric sun and ring rollers on the inner and outer surfaces respectively (see figure 7.39). By applying input to either the sun or ring roller, the unit may act as a speed decreaser or increaser, respectively. Previous traction devices had used only one row of planetary rollers, which severely handicapped their speed and horsepower capacity. Reaction torque is transferred to or from the ring by a pair of rolling-element bearings installed in the outer planetary rollers. This considerably reduces the total number of drive bearings needed, and, together with the self-correcting positioning of the rollers because of their three symmetric contact points with adjacent elements, ensures a quiet and almost vibrationless power transmission with no need for especially fine tolerances. This characteristic is further enhanced by the presence of a thin elastohydrodynamic lubricant film between all contacting surfaces, which, along with the elastic compliance of the rollers, provides effective vibration damping. An automatic loading mechanism is incorporated as part of the latest design to ensure that the correct ratio of traction forces to normal contract forces is maintained under all conditions. This must be carefully optimized to provide minimum slip without unnecessary wear on the roller elements. In general, two planetary rows are suitable for speed ratios less than 35 to 1, and three planetary rows for speed ratios up to 150 to 1.

The successful operation of a 15:1 speed ratio test rig has recently been validated at speeds up to 7300 rpm, power levels up to 170 horsepower, and

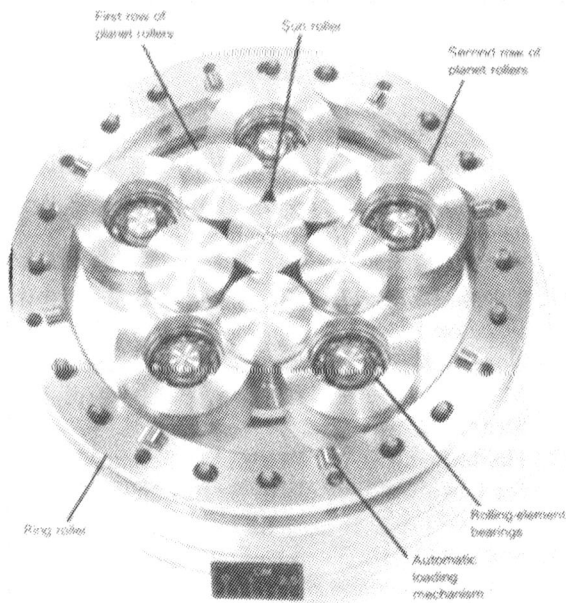

Figure 7.39 - Example of nasvytis traction drive unit (by courtesy of NASA Lewis Research Center).

a peak efficiency of 95%. The overall speed efficiency was 98.6%, which compares favorably with gear-driven systems. Based on other models already built, it is considered that power-to-weight ratios of 4.5 horsepower/lb for continuous operation are achievable, and units of over 500 horsepower capacity can be easily designed. It should also be noted that the test models used a considerable amount of knowledge gained from NASA research programs in traction fluids, rolling element fatigue, bearing steels, and roller kinematics.

These devices have particular application in such areas as the machine tool industry, where the high speed rates would result in much better quality, and in high speed drive systems in gas turbines or turbomachinery generally.[70] Currently, their utility in replacing noisy helicopter main rotor transmissions and aircraft drive systems is being investigated. Although no formal acoustic or vibration tests have yet been conducted on the revised traction drive units, indications are that sound pressure level reductions of at least 10 dB compared to conventional drives are possible. Furthermore, the approximately white noise spectrum which they generate is easier to control than the spectrum associated with toothed-drive systems.

REFERENCES

1. Lighthill, M. J., On Sound Generated Aerodynamically, II., Turbulence as a Source of Sound, Proc. Roy. Soc. (London) ser. A, vol. 222, no. 1148, Feb. 1954, p. 1.
2. Stone, J. R., Interim Prediction Method for Jet Noise, NASA TM X-71618, 1975.
3. Maestrello, L., On the Relationship between Acoustic Energy Density Flux near the Jet Axis and Far-Field Acoustic Intensity, NASA TN D-7269, 1973.
4. Sahlin, S., and R. Langhe, Origins of Punch Press and Air Nozzle Noise, Noise Con. Eng., vol. 3, no. 3, November–December 1974, p. 4.
5. Lord, H. W., H. A. Evensen, and R. J. Stein, Pneumatic Silencer for Exhaust Valves and Parts Ejectors, Sound and Vib., May 1977, p. 26.
6. Huff, R. H., A Simple Noise Suppressor Design for Vented High Pressure Gas, NASA Tech. Brief, summer 1979, p. 278.
7. Huff, R. H., and D. E. Groesbeck, Cold-Flow Acoustic Evaluation of a Small-Scale, Divergent, Lobed Nozzle for Supersonic Jet Noise Suppression, NASA TM X-3210, 1975.
8. Maestrello, L., Initial Results of a Porous Plug Nozzle for Supersonic Jet Noise Suppression, NASA TM 78802, 1978.
9. Maestrello, L., An Experimental Study on Porous Plug Jet Noise Suppressor, AIAA Paper 79-0673, March 1979.
10. Gutierrez, O. A., and J. R. Stone, Developments in Aircraft Jet Noise Technology. Aircraft Safety and Operating Problems, NASA SP-416, 1976, p. 497.
11. Schwartz, I. R., Minimization of Jet and Core Noise by Rotation of Flow, NASA Tech Brief B75-10131, 1975.
12. Reethoff, G., Control Valve and Regulator Noise Generation, Propagation, and Reduction, Noise Con. Eng., vol. 9, no. 2, September–October 1977, p. 74.
13. Cheremisinoff, P. N., and P. P. Cheremisinoff, Industrial Noise Control Handbook, in *Hydrodynamic Control of Valve Noise,* Chapter 14, Ann Arbor Science Publishers, 1977.
14. Industrial Noise Services, Inc., Industrial Noise Control Manual (prepared for NIOSH), available from NTIS as DHEW Pub. No. (NIOSH) 75-183, 1975.

15. Regetz, J. D., Jr., An Experimental Determination of the Dynamic Response of a Long Hydraulic Line, NASA TN D-576, 1960.

16. Blade, R. J., W. Lewis, and J. H. Goodykoontz, Study of a Sinusoidally Perturbed Flow in a Line Including a 90° Elbow with the Flexible Supports, NASA TN D-1216, 1962.

17. Lewis, W., R. J. Blade, and R. G. Dorsch, Study of the Effect of a Closed-End Side Branch on Sinusoidally Perturbed Flow of Liquid in a Line, NASA TN D-1876, 1963.

18. Blade, R. J., and C. M. Holland, Attenuation of Sinusoidal Perturbations in Nonviscous Liquid Flowing in a Long Line with Distributed Leakage, NASA TN D-3563, 1966.

19. Stephenson, J., and H. A. Hassan, The Spectrum of Combustion-Generated Noise, J. Sound Vib., vol. 53, no. 2, 1977, p. 283.

20. Shivashankara, B. N., W. C. Strahle, and L. C. Cho, Combustion Generated Noise in Turbopropulsion Systems, Second Interagency Symposium on University Research in Transportation Noise, Vol. II, June 1974, p. 714.

21. Putnam, A. A., Combustion Noise in Industrial Burners, Noise Con. Eng., vol. 7, no. 1, July–August 1976, p. 24.

22. Plett, E. G., M. D. Leshner, and M. Summerfield, Combustion Geometry and Combustion Roughness Relation to Noise Generation, Second Interagency Symposium on University Research in Transportation Noise, Vol. II, June 1974, p. 723.

23. Plett, E. G., M. D. Leshner, and M. Summerfield, Effect of Flame Distribution on Noise Generation, Third Interagency Symposium on University Research in Transportation Noise, November 1975, p. 207.

24. Mugridge, B. D., Noise Characteristics of Axial and Centrifugal Fans as used in Industry—A Review, Shock and Vibration Digest, Feb. 1979, p. 17.

25. Longhouse, R. E., Vortex Shedding Noise of Low Tip Speed, Axial Flow Fans, J. Sound Vib., vol. 53, no. 1, 1977, p. 25.

26. Mellin, R. C., Selection of Minimum Noise Fans for a Given Pumping Requirement, Noise Con. Eng., vol. 4, no. 1, January–February 1975, p. 35.

27. Graham, J. B., How to Estimate Fan Noise, Sound and Vibration, May 1972, p. 24.

28. Mellin, R. C., Noise and Performance of Automotive Cooling Fans, SAE Tech Paper 800031, Feb. 1980.

29. Beranek, Leo L., ed., *Noise and Vibration Control,* McGraw-Hill Book Co., 1971.

30. Graham, J. B., Fans and Blowers, in *Handbook of Noise Control,* Cyril M. Harris, ed., second ed., Chapter 27, McGraw-Hill Book Co., 1979.

31. Lakshminarayana, B., Influence of Turbulence on Fan Noise, in Workshop on Ventilation System Cooling Fan Noise, Nov. 1975, M. Sevik and J. Pierpoint, eds., Naval Sea Systems Command, 1976, p. 61.

32. Heidmann, M. F., Interim Prediction Method for Fan and Compressor Source Noise, NASA TM X-71763, 1975.

33. Soderman, P. T., Leading Edge Serrations which Reduce the Noise of Low-Speed Rotors, NASA TN D-7371, 1973.

34. Hersh, A. S., and R. E. Hayden, Aerodynamic Sound Radiation from Lifting Surfaces with and without Leading-Edge Serrations, NASA CR-114370, 1971.

35. Hayden, R. E., Some Advances in the Design Techniques for Low Noise Operation of Propellers and Fans, Noise-Con. 77, 1977, p. 381.

36. Hayden, R. E., D. B. Bliss, B. S. Murray, et al., Analysis and Design of a High Tip Speed, Low Source Noise Aircraft Fan Incorporating Swept Leading Edge Rotor and Stator Blades, NASA CR-135092, 1977.

37. Moiseev, N., B. Lakshminarayana, and D. E. Thompson, Noise due to Interaction of Boundary-Layer Turbulence with a Compressor Rotor, J. Aircraft, vol. 15, no. 1, 1978, p. 53.

38. Goldstein, M. E., Highlights 1979, AIAA Journal, vol. 17, no. 12, December 1979.

39. Webb, J. D., ed., *Noise Control in Industry,* Halstead Press, 1976.

40. Bell, L. H., *Fundamentals of Industrial Noise Control,* Second ed., Harmony Publications, 1973.

41. Tyler, J. M., and T. G. Safrin, Axial Flow Compressor Noise Studies, SAE Trans. vol. 70, 1962, pp. 309–332.

42. Lumsdaine, E., and J. Chang, Influence of Blade Characteristics on Axial Flow Compressor Noise, AIAA Paper 76-570, July 1976.

43. Copeland, W., J. L. Crigler, and A. C. Dibble, Jr., Contribution of Downstream Stator to the Interaction Noise of a Single-Stage Axial-Flow Compressor, NASA TN D-3892, 1967.

44. Posey, J. W., High Frequency Sound Attenuation in Short-Flow Ducts, NASA TM 78708, 1978.

45. Atvars, J., and R. A. Mangiarotty, Parametric Studies of the Acoustic Behavior of Duct-Lining Materials, J. Acoust. Soc. Amer., vol. 48, no. 3, 1970, p. 815.

46. Mangiarotty, R. A., Acoustic-Lining Concepts and Materials for Engine Ducts, J. Acoust. Soc. Amer., vol. 48, no. 3, 1969, p. 783.

47. Marsh, A. H., I. Elias, J. C. Hoehne, and R. L. Frasca, A Study of Turbofan-Engine Compressor-Noise-Suppression Techniques, NASA CR-1056, 1968.

48. Hersh, A. S., and B. Walker, Effect of Grazing Flow on the Acoustic Impedance of Helmholtz Resonators Consisting of Single and Clustered Orifices, CR 3177, 1979.

49. Rice, E. J., Spinning Mode Sound Propagation in Ducts with Acoustic Treatment, NASA TN D-7913, 1975.

50. Rice, E. J., Modal Propagation Angles in Ducts with Soft Walls and their Connection with Suppressor Performance, AIAA Paper 79-0624, March 1979.

51. Rice, E. J., Attenuation of Sound in Ducts with Acoustic Treatment—A Generalized Approximate Equation, NASA TM X-71830, 1975.

52. Zorumski, W. E., and J. P. Mason, Multiple Eigenvalues of Sound-Absorbing Circular and Annular Ducts, J. Acoust. Soc. Amer., vol. 55, 1974, p. 1158.

53. Minner, G. L., and E. J. Rice, Annular Acoustic Liners for Turbofan Engines, NASA Tech. Brief, Spring 1979, p. 132.

54. Kraft, R. W., J. E. Pass, and L. R. Clark, Effects of Multi-Element Acoustic Treatment on Compressor Inlet Noise, AIAA Paper 76-515, July 1976.

55. Chestnutt, D., and C. E. Feiler, Advanced Inlet Duct Noise Reduction Concepts. Aircraft Safety and Operating Problems, NASA SP-416, 1976, p. 481.

56. Zorumski, W. E., Acoustic Theory of Axisymetric Multisectioned Ducts, NASA TR R-419, 1974.

57. Zorumski, W. E., Noise Suppressor, U.S. Patent No. 3 830 335, August 1974.

58. Kraft, R. E., R. E. Motsinger, W. H. Gauden, and J. F. Link, Analysis Design, and Test of Acoustic Treatment in a Laboratory Inlet Duct, NASA CR-3161, July 1979.

59. ASHRAE Handbook and Product Directory, Systems Volume, Chapter 35, Amer. Soc. Heat, Refrig., and Air-Con. Eng., New York, 1976.

60. Soderman, P. T., and L. E. Hoglund, Wind-Tunnel Fan Noise Reduction including Effects of Turning Vanes on Noise Propagation, AIAA Paper 79-0642, March 1979.

61. Lansing, D. L., Exact Solution for Radiation of Sound from a Semi-Infinite Circular Duct with Application to Fan and Compressor Noise. Analytic Methods in Aircraft Aerodynamics, NASA SP-228, October 1969, p. 323.

62. Scharton, T. D., and M. D. Sneddon, Acoustical Properties of Materials and Muffler Configurations for the 80 × 120 Foot Wind Tunnel, NASA CR-152065, 1977.

63. Ingard, U., and W. P. Patrick, A New Approach to Acoustic Filtering with Lined Ducts, Third Interagency Symposium on University Research in Transportation Noise, November 1975, p. 522.

64. Parrott, T. L., An Improved Method for the Design of Expansion-Chamber Mufflers with Application to an Operational Helicopter, Cosmic Program Abstract, LAR-11548, 1974.

65. Anyos, T., L. Christy, R. Lizak, and J. Wilhelm, Technology Transfer-Transportation, (prepared for NASA by SRI International), Annual Report, Contract NAS2-9848, 1978, p. 39.

66. Hart, F. D., D. I. Neal, and F. O. Smetana, Industrial Noise Control: Some Case Histories, Volume 1, NASA CR-142314, 1974.

67. Giurgiuman-Negrea, H., M. Creto, and I. Moraru, Aspects of Reducing Spur Gear Noise, The Fourth National Conference on Acoustics, Vol 1A: Noise and Vibrations

Control, NASA TT F-15, 385, June 1974, p. 124.

68. Schlegel, R. Y., R. J. King, and H. R. Mull, How to Reduce Gear Noise, Machine Design, Feb. 27, 1964, pp. 134–142.

69. Loewenthal, S. H., N. E. Anderson, and A. L. Nasvytis, Performance of a Nasvytis Multiroller Traction Drive, NASA TP 1378/AVRADCOM TR 78-36, 1978.

70. NASA Activities, National Aeronautics and Space Administration, Vol. 10, no. 10, 1979, p. 8.

APPENDIX

OSHA Noise Regulations and Assessment of Hearing Impairment

A. OSHA STANDARD

Under the Occupational Safety and Health Act of 1970, two sets of regulations were promulgated in August 1971 pertaining to general industry[1] and to the construction field.[2] The terms of these are essentially the same and provide as follows:

"When employees are subjected to sound levels exceeding those listed in Table G-16, feasible administrative or engineering controls shall be utilized. If such controls fail to reduce sound levels within the levels of Table G-16, personal protective equipment shall be provided and used to reduce sound levels within the levels of the Table.

Duration per Day (hours)	Sound Level dB (A) Slow Response
8	90
6	92
4	95
3	97
2	100
1½	102
1	105
½	110
¼ or less	115

TABLE G-16–Permissible Noise Exposure

"When the daily noise exposure is composed of two or more periods of noise exposure of different levels, their combined effect should be considered, rather than the individual effect of each. If the sum of the following fractions: $C_1 / T_1 + C_2 / T_2 + \ldots + C_n / T_n$ exceeds unity, then the mixed exposure should be considered to exceed the limit value. T_n indicates the total time of exposure permitted at the level n, and C_n is the actual time of exposure at that level.

"Exposure to impulsive or impact noise should not exceed 140 dB peak sound pressure level."

To achieve compliance with the standard, it is necessary first to compute the total noise exposure of each employee and then, if necessary, examine the permutations of noise level and exposure duration which will ensure a daily noise dose of less than unity. Even with levels greater than 90 dB(A), the total exposure may be within the standard. Conversely, even when compliance is achieved for each individual level, the total noise exposure for an 8-hour period may be greater than unity. Both of these points are illustrated in the following example with the modified noise level series 1 and 2, respectively.

Employee exposure condition	Noise level (dB (A)) Initial	Modified 1	Modified 2	Actual duration, C (hr)	Permitted duration (hr) Initial	Modified 1	Modified 2
1	95	83	88	2 7	4 0	—	—
2	102	94	95	3 6	1 5	4 4	4
3	99	88	88	0 2	2 3	—	—
4	108	91	100	1 5	0 7	6 9	2

	Daily noise dose		
	Initial	Modified 1	Modified 2
	0.7	—	—
	2.4	0.8	0.9
	0.1	—	—
	2.1	0.2	0.8
Totals	5.3	1.0	1.7

If the daily noise exposure exceeds unity, it is necessary to calculate the noise reduction necessary to achieve compliance with the standard. It has been established that the question of "feasibility" of noise control measures rests on both economic and technical factors.[3] In some noise control problems it may not be possible to establish a total cost estimated until several stepwise stages of engineering control have been implemented. Cost estimates should be compared with the general level of total expenditure per employee to meet OSHA compliance for that particular industry.[4]

In October 1974, proposed changes in the OSHA noise standard were published.[5] Although there was to be no alteration of the basic noise exposure limits, it was proposed that hearing protectors be worn in exposure conditions of above 85 dB(A) and that monitoring audiometric testing programs be established. At present there are no standard procedures for using audiometry as part of a hearing conservation program.

B. HEARING IMPAIRMENT

There is considerable debate as to what constitutes hearing impairment as reflected in different state procedures for assessing compensation, although the most common criterion invoked is that of the American Academy of Ophthalmology and Otolaryngology (AAOO). This uses a hearing threshold of 25 dB relative to audiometric zero, averaged over 500, 1000, and 2000 Hz. However, there is no question that considerable noise-induced hearing loss occurs at 3000 and 4000 Hz, and there is strong evidence that hearing ability at these frequencies is very important under less than ideal listening conditions. Figure A.1 shows the range of hearing loss risks in workers after 20 years of continuous employment.[6] The lower curve shows the risk of permanent impairment after all factors except workplace noise have been eliminated; the upper curve includes the effects of disease, abnormalities, and intense noise experienced outside work. A monitoring audiometry program consists of an initial aural examination of all new company employees and their audiometric testing at regular intervals thereafter. Apart from benefiting the health of employees, such a program also enables the company to check on any noise-induced hearing loss caused by previous workplace exposure and to identify individuals whose hearing is already seriously diminished.

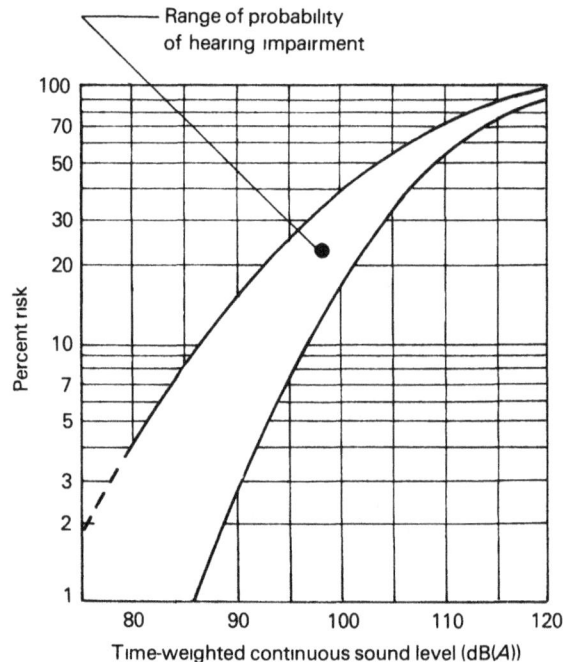

Figure A.1 – Probability of workers having a hearing threshold of 25 dB or more, averaged over 500, 1000, and 2000 Hz, after 20 years of exposure to workplace noise. (Used by permission of Acoustical Publications.)

REFERENCES

1. Anon., 29 Code of Federal Regulations 1910. 95(b)(1). Federal Register, vol. 36, May 29, 1971, p. 10466.

2. Anon., 29 Code of Federal Regulations 1926. 52(b).

3. Parashes, P. T., A Legal Overview of the OSHA Noise Standard at 29 CFR 1910. 95(b)(1), Sound and Vibration, June 1978, p. 10.

4. Heggie, A. S., The OSHA Noise Standard—How to Live With It, Sound And Vibration, June 1978, p. 20.

5. Occupational Noise Exposure—Proposed Requirements and Procedures, Federal Register, vol. 39, 207, p. 20.

6. Broce, R. D., et al., Workplace Noise Exposure Control—What are the Costs and Benefits?, Sound and Vibration, September 1976, p. 12.

Glossary

Units

A Total room absorption, equal to the sum of the products of the area of each ith surface element, S_i, and the corresponding absorption coefficients, α_i, defined mathematically as

$$A = \frac{\Sigma S_i \alpha_i}{i}$$

metric sabins m²

B_y Stiffness (Young's) modulus, the ratio of stress to strain in a solid bar   N/m²

D Acoustic energy density contained in a sound field   J/m³

D Diameter   m

I Acoustic intensity, the average rate of flow of sound energy through a unit area perpendicular to the direction under consideration   W/m²

I_{ref} Reference sound intensity taken as 10^{-12} watts per square meter   W/m²

IL   Acoustic intensity level expressed in decibel notation as 10 times the logarithm (base 10) of the ratio of a sound intensity to the reference sound intensity, 10^{-12} W/m²   dB

InL   Insertion loss, the difference in sound pressure level at a particular location before and after acoustic treatment is applied   dB

K Number of blades, gear teeth, etc.

L Length (cavity, muffler, etc.)   m

L_{dn} Day–night average sound level, the 24-hour equivalent continuous sound level with a 10-dB weighting applied to sound levels between 10:00 p.m. and 7:00 a.m.   dB(A)

L_{eq} Equivalent continuous sound level, the level of a steady sound which, over the same time period, contains the same sound energy as the time-varying sound   dB(A)

L_x x-Percentile exceeded sound level, the level (fast) equaled or exceeded by a time-varying sound x percent of a given time period   dB(A)

129

M — Mach number, the ratio of the flow velocity to the ambient speed of sound

N — Fresnel number, defined mathematically as

$$N = \pm \frac{2\delta}{\lambda}$$

where
δ = path length difference around barrier
λ = wavelength of interest

N — Shaft or fan rotational speed — rpm, rps

NR — Noise reduction, the difference between two sound pressure levels across a partition — dB

OASPL — Overall sound pressure level, i.e., the sound pressure level over all frequency bands in the audio frequency range (usually 31–16 000 Hz) — dB

P_T — Total pressure — N/m² (Pa)

P_o — Reference ambient atmospheric pressure, normally taken as equal to 1.012×10^2 newtons per square meter at sea level — N/m²

PSD — Power spectral density, defined mathematically as

$$\text{PSD} = \lim_{B_l \to 0} \frac{W_l}{B_l}$$

i.e., in the limit as the bandwidth approaches zero, the PSD is the ratio of power in a bandwidth W_l, divided by that bandwidth B_l — W/Hz

PWL — Sound power level expressed in decibel notation as 10 times the logarithm (base 10) of the ratio of a sound power to the reference sound power, 10^{-12} watts — dB

Q — Air flow rate — m³/min

Q — Directivity factor, the ratio of sound intensity at angle θ to the intensity at the same location when the source is radiating omnidirectionally

SPL — Sound pressure level expressed in decibel notation as 20 times the logarithm (base 10) of the ratio of a sound pressure to the reference sound pressure, 2×10^{-5} N/m² — dB

R	Sound transmission loss of an insulating material, defined as 10 times the logarithm (base 10) of the reciprocal of the transmission coefficient	dB

R_c Room constant, a convenient measure of room absorption; defined mathematically as

$$R_c = \frac{S\bar{\alpha}}{1 - \bar{\alpha}}$$

where   metric
S = total room internal area, m^2 sabins,
$\bar{\alpha}$ = average absorption coefficient of room surfaces   m^2

RMS   Root mean square

S Area (partition area, hypothetical enveloping surface, etc.)   m^2

T Absolute temperature   K

T Period, equal to the time taken for a regular waveform to complete one cycle of oscillation   min, s

T_R Reverberation time according to Sabine, defined mathematically as

$$T_R = \frac{0.161V}{A}$$

where
V = room volume, m^3
A = room absorption, metric sabins   s

U Acoustic medium particle velocity in direction of propagation   m/s

V Volume (room, cavity, etc.)   m^3

W Acoustic power, the total sound energy produced by a source in unit time   W

W_{ref} Reference sound power taken as 10^{-12} watts   W

Z Specific acoustic impedance, the ratio of complex acoustic pressure to complex particle velocity at the surface of a material   Ns/m^3 (mks rayls)

c Velocity of sound propagation in air, normally taken as 343 meters per second at room temperature and pressure   m/s

dB   10 times the logarithm (base 10) of the ratio of two physical quantities   dB

dB(A)	A-Weighted decibels, normally summed across the entire audio frequency range	dB(A)
dB(C)	C-Weighted decibels, normally summed across the entire audio frequency range	dB(C)
d	Separation distance of two surfaces	cm, m
f	Frequency, the number of complete oscillations of a periodic waveform that occur in one second	Hz
p	Acoustic sound pressure, the difference between the actual total pressure and the mean ambient pressure	N/m^2
p_{ref}	Reference sound pressure taken as 2×10^{-5} newtons per square meter	N/m^2
p_{rms}	Root mean square of instantaneous sound pressure over a specified time interval	N/m^2
r	Distance from noise source	m
r_c	Distance from source at which transition from direct to reverberant fields begins to occur	m
v	Air or liquid flow velocity	m/s
x	Distance coordinate	m
α	Absorption coefficient, taken usually as the fraction of random incidence sound power absorbed by a material	
δ	Path length difference, the difference between direct path from source to receiver and the shortest path around a barrier	m
ζ	Damping ratio, the ratio of the damping coefficient of the material under consideration to the critical damping coefficient	
η	Damping loss factor, equal to twice the ratio of the material damping coefficient to the critical damping coefficient of that material	
λ	Wavelength, equal to the distance between congruent points on successive oscillations of a single frequency tone	m
ϱ	Density	kg/m^3
ϱc	Characteristic acoustic impedance, the real part of the specific acoustic impedance Z, and equal to this only for free progressive plane waves	Ns/m (mks rayls)
τ	Transmission coefficient, taken as the fraction of incident sound power transmitted through a material	

SUBSCRIPTS

E	Enclosure variable
REV	Reverberant field variable
S	Component of variable passing through area S
T	Total value
j	Jet variable
n	Normal component of variable or nth harmonic component
o	Equilibrium or ambient value of variable
p	Peak value of variable
ref	Reference value of variable
res	Value of variable at resonance or at undamped natural frequency of a loaded system
rms	Root mean square
t	Test value of variable
θ	Directional component at angle θ of variable

Bibliography

GENERAL TEXTS

AIHA, *Industrial Noise Manual*, second ed., American Industrial Hygiene Assoc., 1966.

Beranek, Leo L., ed., *Noise and Vibration Control*, McGraw-Hill Book Co., 1971.

Berendt, R. D., E. L. R. Corliss, and M. S. Ojalvo, *Quieting: A Practical Guide to Noise Control*, National Bureau of Standards Handbook 119, U.S. Dept. of Commerce, 1976.

Broch, J. T., *Acoustic Noise Measurements,* second ed., Bruel & Kjaer, 1971.

Bell, L. H., *Fundamentals of Industrial Noise Control*, second ed., Harmony Publications, 1973.

Cheremisinoff, P. N. and P. P. Cheremisinoff, *Industrial Noise Control Handbook*, Ann Arbor Science Publishers, 1977.

Cremer, L., and M. Heckl (E. E. Ungar, transl.), *Structure-Borne Sound*, Springler-Verlag, 1973.

Crocker, M. J., ed., *Reduction of Machinery Noise*, Revised ed., Purdue University, 1975.

Diehl, G. M., *Machinery Acoustics*, John Wiley & Sons, 1973.

Faulkner, L. L., ed., *Handbook of Industrial Noise Control*, Industrial Press, 1976.

Harris, C. M., ed., *Handbook of Industrial Noise Control*, Second ed., McGraw-Hill Book Co., 1979.

Lyon, R. H., *Statistical Energy Analysis of Dynamical Systems: Theory and Applications*, MIT Press, 1975.

Miller, R. K., and W. V. Montone, *Handbook of Acoustical Enclosures and Barriers*, Fairmont Press, 1978.

Peterson, A. P. G., and E. E. Gross, Jr., *Handbook of Noise Measurement*, Seventh ed., General Radio, 1974.

Salmon, V., J. S. Mills, and A. C. Peterson, *Industrial Noise Control Manual* (prepared for NIOSH), Dept. of Health, Education, and Welfare, 1975.

Webb, J. D., ed., *Noise Control in Industry*, Sound Research Laboratories, Halstead Press, 1976.

Yerges, L. F., *Sound, Noise, and Vibration Control*, Second ed., Van Nostrand Reinhold Co., 1978.

JOURNALS AND PERIODICALS

Acustica: S. Hirze, Stuttgart 1, Birkenwaldstr. 44, Postfach 347, Germany (monthly).

Applied Acoustics: Applied Science Publishers, Ltd., Ripple Rd., Barking, Essex, England (quarterly).

Journal of Sound and Vibration: Academic Press, Inc., 111 Fifth Avenue, New York, N.Y. 10003 (bi-weekly).

Noise Control Engineering: Institute of Noise Control Engineering, P.O. Box 3206, Arlington Branch, Poughkeepsie, N.Y. 12603 (bi-monthly).

Noise/News: Institute of Noise Control Engineering, P.O. Box 3206, Poughkeepsie, N.Y. 12603 (bi-monthly).

Sound and Vibration: Acoustical Publications, Inc., 27101 E. Oviatt Road, Bay Village, Ohio 44140 (monthly).

The Journal of the Acoustical Society of America: Acoustical Society of America, 335 E. 45 Street, New York, N.Y. 10017 (monthly).

The Shock and Vibration Digest: The Shock and Vibration Information Center, Code 8404 Naval Research Laboratory, Washington, D.C. 20375 (monthly).

NOISE CONFERENCE PROCEEDINGS

Interagency Symposium on University Research in Transportation Noise: Proceedings of 1st at Stanford University, 1973; 2nd at North Carolina State University, 1974; 3rd at University of Utah, 1975.

International Congress on Acoustics: 1st–9th Congress Proceedings, held every 3 years. Elsevier Pub. Co., and Halstead Press.

Inter-Noise: Proceedings of the International Conference on Noise Control Engineering.

Noise Control Foundation, P.O. Box 3469, Arlington Branch, Poughkeepsie, N.Y. 12603, 1972.

Noise and Vibration Control Engineering: *Proceedings of the Purdue Noise Control Conference*. Purdue University, Lafayette, Indiana, July 14–16, 1971.

Noise-Con: *Proceedings of the National Conference on Noise Control Engineering*. Noise Control Foundation, P.O. Box 3469, Arlington Branch, Poughkeepsie, N.Y. 12603.

Noisexpo: *Proceedings of the Technical Program*: *National Noise and Vibration Control Conference*, Sound and Vibration, 27101 E. Oviatt Road, Bay Village, Ohio, 44140.

The Fourth National Conference on Acoustics (Rumania): Vol. I, A. Noise and Vibration Control, 1973. NASA TT F-15, 375, 1974.

The Shock and Vibration Bulletin: Proceedings of the Annual Shock and Vibration Symposium. The Shock and Vibration Information Center, Code 8404 Naval Research Laboratory, Washington, D.C. 20375.

University Noise Research: Proceedings of the EPA–University Noise Seminar. EPA 550/9-77 Office of Noise Abatement and Control, EPA, Washington, D.C. 20460, October 18–20, 1976.

GENERAL POLICY AND INFORMATION DOCUMENTS

Center for Policy Alternatives, MIT: Some Considerations in Choosing an Occupational Noise Exposure Regulation. EPA 550/9-76-007. Office of Noise Abatement and Control, U.S. Environmental Protection Agency, Washington, D.C., 1976.

EPA: Toward a National Strategy for Noise Control. Office of Noise Abatement and Control, U.S. Environmental Protection Agency, Washington, D.C., 1977.

Kroes, P., R. Fleming, and B. Lempert, List of Personal Hearing Protectors and Attenuation Data. HEW Pub. No. (NIOSH) 76-120. National Institute for Occupational Safety and Health, U.S. Dept. of Health, Education, and Welfare, 1975.

L. S. Goodfriend Associates: Noise from Industrial Plants. NTID 300.2. Office of Noise Abatement and Control, U.S. Environmental Protection Agency, Washington, D.C., 1971.

NIOSH: Criteria for a Recommended Standard . . . Occupational Exposure to Noise. HSM 73-11001. National Institute for Occupational Safety and Health, U.S. Dept. of Health, Education, and Welfare, 1972.

Schemidek, M. E., et al., Survey of Hearing Conservation Programs in Industry. HEW Pub. No. (NIOSH) 75-178. National Institute for Occupational Safety and Health, U.S. Dept. of Health, Education, and Welfare, 1975.

Wyle Laboratories: Community Noise. NTID 300.3. U.S. Environmental Protection Agency, Washington, D.C. 1971.

MANUALS

Emerson, P. D., J. R. Bailey, and F. D. Hart, *Manual of Textile Industry Noise Control*, North Carolina State University, 1978.

Hart, F. D., C. L. Neal, and F. O. Smetana, *Industrial Noise Control: Some Case Histories*, Vol. 1, NASA CR 142314, North Carolina Science and Technology Research Center, 1974.

Pearsons, K. S., and R. Bennett, *Handbook of Noise Ratings*, NASA CR-2376, 1974.

Purcell, W. E., and B. L. Lempert, *Compendium of Materials for Noise Control*, HEW Pub. No. (NIOSH) 75-165, NIOSH, U.S. Dept. Health, Education, and Welfare, 1975.

Salmon, V., et al., *Industrial Noise Control Manual* (revised), HEW Pub. No. (NIOSH) 79-117, prepared by Industrial Noise Services, Inc. for NIOSH, U.S. Dept. Health, Education, and Welfare, 1979.

Simpson, M. A., *Noise Barrier Design Handbook*, FHWA-RD-76-58, prepared by Bolt Beranek and Newman, Inc., for Federal Highway Administration, U.S. Dept. of Commerce, 1976.

Wyle Laboratories, *Aerodynamically Quiet Circular Saw Blades . . . A Design and Application Manual*, prepared by Wyle Labs., for Woodworking Machinery Manufacturers of America, Philadelphia, Pa., 1978.

BIBLIOGRAPHIES AND RESEARCH REVIEWS

EPA: Noise Technology Research Needs and The Relative Roles of the Federal Government and The Private Sector, EPA 550-/9-79-311, Office of Noise Abatement and Control, U.S. Environmental Protection Agency, 1978.

Pusey, H. C., and R. H. Bolin, An International Survey of Shock and Vibration Technology, Shock and Vibration Information Center, Naval Research Laboratory, Washington, D.C., 1979.

The Federal Interagency Machinery and Construction Noise Research Panel: Federal Research, Development and Demonstration Programs in Machinery and Construction Noise, EPA 550/9-78-306, Office of Noise Abatement and Control, U.S. Environmental Protection Agency, 1978.

The Federal Interagency Surface Transportation Noise Research Panel: Federal Research, Development and Demonstration Programs in Surface Transportation Noise, EPA 550/9-78-305, Office of Noise Abatement and Control, U.S. Environmental Protection Agency, 1978.

The Technology Application Center, Institute for Social Research and Development: Noise Pollution Resources Compendium, TAC Bibliographic Series No. 2, The University of New Mexico, 1972.

Shock and Vibration Information Center: An International Survey of Shock and Vibration Technology, Naval Research Laboratory, Washington, D.C., 1979.

NASA TECHNICAL PAPERS

Clemons, A., Quiet Clean Short-Haul Experimental Engine (QCSEE) Acoustic Treatment Development and Design, NASA CR-135266, 1979.

Copeland, W. Latham, Inlet Noise Studies for an Axial-Flow Single-Stage Compressor, NASA TN D-2615, 1965.

Getline, G. L., Low-Frequency Noise Reduction of Lightweight Airframe Structures, NASA CR-145104, 1976.

Goldstein, M. E., Aeroacoustics, NASA SP-346, 1974.

Goldstein, M. E., and W. L. Howes, New Aspects of Subsonic Aerodynamic Noise Theory, NASA TN D-7158, 1973.

Hersh, A. S., and R. E. Hayden, Aerodynamic Sound Radiation from Lifting Surfaces with and without Leading-Edge Serrations, NASA CR-114370, 1971.

Howes, W. L., Overall Loudness of Steady Sounds According to Theory and Experiment, NASA RP 1001, 1979.

Howlett, J. T., S. A. Clevenson, J. A. Rupf, and W. J. Snyder, Interior Noise Reduction in a Large Civil Helicopter, NASA TN D-8477, 1977.

Lambert, R. F., Acoustical Properties of Highly Porous Fibrous Materials, NASA TM 80135, 1979.

Levine, L. S., and J. J. DeFelice, Civil Helicopter Research Aircraft Internal Noise Prediction, NASA CR-145146, 1977.

Manning, J. E., and P. J. Remington, Statistical Energy Methods, NASA CR-118987, 1971.

McDonald, W. B., R. Vaicaitis, and M. K. Myers, Noise Transmission Through Plates into an Enclosure, NASA TP 1173, 1978.

Montegani, F. J., Some Propulsion System Noise Data Handling Conventions and Computer Programs used at the Lewis Research Center, NASA TM X-3013, 1974.

Plumblee, H. E., Jr., P. D. Dean, G. A. Wynne, and R. H. Burrin, Sound Propagation in and Radiation from Acoustically Lined Flow Ducts: A Comparison of Experiment and Theory, NASA CR-2306, 1973.

Parrott, R. L., An Improved Method for the Design of Expansion-Chamber Mufflers with Application to an Operational Helicopter, NASA TN D-7309, 1973.

Smith, P. W., and R. H. Lyon, Sound and Structural Vibration, NASA CR-160, 1965.

Stone, J. R., Interim Prediction Method for Jet Noise, NASA TM X-71618, 1975.

Vaicaitis, R., Noise Transmission by Viscoelastic Sandwich Panels, NASA TN D-8516, 1977.

Zorumski, W. E., Acoustic Theory of Axisymmetric Multisectioned Ducts, NASA TR R-419, 1974.